Integrated Planting and Breeding Technologies in Paddy Field

稻田综合种养技术

赵艳岭　王强盛◎主编

中国财富出版社有限公司

图书在版编目（CIP）数据

稻田综合种养技术 = Integrated Planting and Breeding Technologies in Paddy Field : 英文 / 赵艳岭，王强盛主编 . — 北京：中国财富出版社有限公司，2023.12

ISBN 978-7-5047-8047-8

Ⅰ. ①稻… Ⅱ. ①赵… ②王… Ⅲ. ①水稻栽培—英文 Ⅳ. ① S511

中国国家版本馆 CIP 数据核字（2023）第 238062 号

策划编辑 刘静雯　　**责任编辑** 刘静雯　　**版权编辑** 李　洋
责任印制 荀　宁　　**责任校对** 张营营　　**责任发行** 敬　东

出版发行 中国财富出版社有限公司
社　　址 北京市丰台区南四环西路188号5区20楼　　**邮政编码** 100070
电　　话 010-52227588 转 2098（发行部）　　010-52227588 转 321（总编室）
010-52227566（24小时读者服务）　　010-52227588 转 305（质检部）
网　　址 http: //www. cfpress. com. cn　　**排　　版** 宝蕾元
经　　销 新华书店　　**印　　刷** 北京九州迅驰传媒文化有限公司
书　　号 ISBN 978-7-5047-8047-8/S · 0062
开　　本 710mm × 1000mm　1 /16　　**版　　次** 2024年12月第 1 版
印　　张 15.75　　**印　　次** 2024年12月第 1 次印刷
字　　数 274千字　　**定　　价** 49.80 元

编写人员名单

主　编：赵艳岭　王强盛

副主编：李袭杰　余海波

参　编：童华东　梁明华　杨　光　童　星　黄　琼　王　勋
史培华　张新城　张　国　窦　志　李玉祥　朱国兵
黄　帅　刘　麟　刘叶琼　王　姗　张建祥　殷从飞
郑雪娇　谭筱玉　张　莹　史红林　蔡善亚

Forward

Integrated farming of rice refers to the engineering transformation of paddy fields to create a symbiotic and rotational mutual promotion system between rice and aquatic animals. Through large-scale development, industrialized management, standardized production, and brand-oriented operations, this model can achieve stable rice production, increased aquatic product yields, improved economic benefits, and a significant reduction in the use of pesticides and fertilizers. It is an ecological and circular agricultural development model with multiple functions, including stable grain production, fishery promotion, quality improvement, efficiency enhancement, ecology, and environmental protection. From an ecological perspective, it forms a symbiotic ecosystem dominated by rice and aquatic animals (fish, *Trionyx*, shrimp, loaches, crabs) through the introduction of these species into the paddy field ecosystem.

China has a long history of integrated rice-fish farming. However, it was not until the late 1970s that integrated rice-fish farming evolved into a fishery industry. Since then, the development of integrated rice-fish farming has gone through five stages: the initial development period (1978–1985), the production-oriented period (1986–1992), the expansion period (1993–2002), the comprehensive development period (2003–2012), and the comprehensive innovation period (after 2013). Although the area of integrated rice-fish farming has fluctuated over time, the yield of fishery products has continued to increase. Currently, in the process of transforming agricultural methods and adjusting structures, integrated farming of rice in paddy fields has gained significant attention as an efficient, resource-saving, and environmentally friendly ecological and green agricultural production method. Under the strong promotion of the Ministry of Agriculture and Rural Affairs and the active exploration and practice of local governments, competent authorities, scientific and technical personnel, and a large number of farmers, Integrated farming of rice in paddy fields has achieved remarkable results. A large number of typical models of integrated rice-fish farming, centered on rice and dominated by special aquatic products, with characteristics of industrialized management, large-scale development, and standardized production, have emerged continuously. These models have gradually established a technical pattern of integrated farming of rice that "uses fish to promote rice, improves quality and efficiency, protects the environment, and ensures fishery income". This has sparked a new wave of development across the country.

This textbook consists of six chapters. The first chapter, third chapter, and the overall revision and proofreading were completed by Zhao Yanling. The second chapter was completed by Wang Qiangsheng and others. The fourth chapter was completed by Li Xijie, Tong Huadong, Tong Xing, and others. The fifth chapter was completed by Yu Haibo, Liang Minghua, and others. The sixth chapter was completed by Yu Haibo, Yang Guang, and others.

Chapter 1 provides an overview of the machine transplanting technique for integrated farming of high-quality rice in paddy fields. Chapter 2 focuses on integrated farming modes and key techniques in paddy fields, such as integrated rice-shrimp, rice-duck, and rice-crayfish farming techniques. Chapter 3 focuses on the high-yield culture techniques for high-quality rice machine transplanting, detailing the entire production process from base construction, variety selection, and field management to harvesting. This chapter also includes the use and maintenance of agricultural machinery, systematically introducing the structure, application scope, and usage methods of high-performance transplanters. Chapter 4 discusses the green prevention and control technology of diseases and pests of machine-transplanted rice. Chapter 5 analyzes the yields and benefits of integrated farming mode in paddy fields, based on national and regional production practices and multiple case studies. Chapter 6 focuses on the extension and application of technology, elucidating the necessity and feasibility of integrated farming mode in paddy fields from policy orientation, production and social demands, and promotion methods, as well as potential issues and solutions in the promotion process.

The high-quality rice mechanized transplanting cultivation technology for integrated rice in paddy fields is a new technology that is being vigorously promoted. Its technical system is still expanding and improving, and the related disciplinary theories are also gradually enriching and deepening. This technology involves multiple fields such as planting, breeding, and mechanization, with strong interdisciplinary characteristics. It requires the comprehensive application of multidisciplinary knowledge to solve practical problems in production. Due to the limitations of the editors' knowledge and practical experience, there may be some errors and omissions. We sincerely hope that readers and practitioners will provide valuable opinions and suggestions to jointly promote the improvement of this technology to better serve agricultural production practices. During the compilation of this textbook, we referred to a large number of literature materials from many experts and scholars in the relevant fields. We extend our sincere thanks to all the authors.

Editors
Jurong, Jiangsu
June 2024

Contents

Chapter I Overview of Machine Transplanting Technique for Integrated Farming of High-quality Rice in Paddy Fields ·············· 1

Chapter II Integrated Farming Modes and Key Techniques in Paddy Fields ··· 7

Section I Integrated Rice-Duck Farming Technique ····························· 7

Section II Integrated Rice-Crayfish Farming Technique ······················ 23

Section III Integrated Rice-Crab Farming Technique ························· 35

Section IV Integrated Rice-Loach Farming Technique ······················· 41

Section V Restrictions and Countermeasures on the Development of Integrated Rice-Fishery Farming Technique ························· 46

Chapter III High-yield Culture Techniques for High-quality Rice Machine Transplanting ··· 52

Section I Growth Characteristics and Culture Characteristics of Machine-transplanted Rice ·· 52

Section II Seedling Raising Technique for Machine-transplanted Rice ······ 56

Section III Field Transplanting Technique for Machine-transplanted Rice ··· 83

Section IV Water and Fertilizer Management of Machine-transplanting Paddy Fields ·· 116

Chapter IV Green Prevention and Control Technology for Diseases and Pests of Machine-transplanted Rice ························ 141

Section I Occurrence and Identification of Main Diseases in Machine-transplanted Rice ··································· 141

Section II Green Prevention and Control Technology for Integrated Farming in Paddy Fields ··· 199

Chapter V Yield and Benefit Analysis of Integrated Farming Mode in Paddy Fields ··· 209

Section I Economic Benefit ··· 210

Section II Social and Economic Benefits ··· 217

Chapter VI Extension and Application of Technology ··· 221

Section I Overview of Agricultural Science and Technology Extension ··· 221

Section II Agricultural Extension and Operation Services ··· 229

Section III Promotion Cases of Integrated Rice-Fishery Farming ··· 235

Chapter I Overview of Machine Transplanting Technique for Integrated Farming of High-quality Rice in Paddy Fields

I. Development Status and Trend of Integrated Farming in Paddy Fields

1. Origin of the mode of integrated farming in paddy fields

In recent years, with the rapid development of the social economy and the continuous improvement of people's living standards in China, the requirements for food safety and quality have become higher and higher, and the demand for high-quality rice, livestock, poultry, and aquatic products is also increasing. Accordingly, the requirements for agricultural production are also getting higher and higher. Therefore, integrated farming in paddy fields came into being, and after the continuous practice and exploration of farmers and technicians, new production modes and special techniques were formed. Through the summary and improvement of agricultural experts, a systematic theory of integrated farming in paddy fields was formed. Specifically, integrated farming in paddy fields caters to the development demands of society. Through the symbiotic and complementary ecological agricultural farming mode of rice, livestock, poultry, and aquatic products, it can achieve the goal of promoting breeding with planting, complementing planting with breeding, comprehensive utilization, and circular development. This mode changes the traditional single production mode in the past, increases the utilization rate of lands, and greatly improves the ecological environment of farmland. Integrated farming in paddy fields cuts off the effective transmission of harmful microorganisms, which not only reduces the prevalence of diseases and weeds in the farming of plants and animals but also reduces the usage of chemical fertilizers and pesticides to effectively control rural non-point source pollution. Meanwhile, it saves energy and resources, effectively promotes the development of ecological circular agriculture, and realizes the good effects of multiple uses of one field, multiple uses of one water source, and multiple harvests in one season. It can be said that the integrated farming mode in paddy fields is an environment-friendly production mode spawned by the concept of green development in the new era.

It should be said that the combination of planting and breeding in paddy fields is a traditional ecological agricultural mode in China, rather than a creation of modern agriculture. According to the CCTV program "Big Country Granary—China's Rice Bowl", Chinese farmers started farming in paddy fields as early as the Qin and Han Dynasties in agriculture civilization. Some scholars believe that fish farming in paddy fields first appeared in the Han Dynasty and has a history of over 2,000 years. The traditional production mode of fish farming in paddy fields still exists in the terraced areas of Yunnan and Jiangxi Provinces, but a special theoretical concept has not been formed.

2. Concept and connotation

Integrated farming in paddy fields is an efficient stereoscopic farming mode that combines the biological and ecological characteristics of rice with the characteristics of ducks, crayfishes, and fishes, and uses the principle of ecological symbiosis to organically combine rice planting and characteristic breeding under the premise of ensuring the dominant position of rice planting. It can stabilize the grain planting area, improve the ecological environment of paddy fields, and control the growth of pests and weeds in the field with the ducks, fishes, crayfishes, etc. In addition, the excrement of ducks, fishes, and other animals can also provide nutrients for the growth of rice and restore and maintain soil fertility. It can produce good economic benefits and is a cyclic, efficient, safe, and environmentally friendly production mode. At present, the common integrated farming modes in China mainly include duck farming, crayfish farming, fish farming, crab farming, and loach farming in paddy fields.

3. Domestic development status

According to the *Development Report of Integrated Rice-Fishery Farming Industry in China (2018)*, in 2017, there were 27 provinces in China that implemented integrated rice-fishery farming, with an increasing area of more than 1,333,300 hm^2, and an output value of more than CNY 1,000 was created per mu (1 mu≈667 m^2). Yunnan Province, Jiangxi Province, Hunan Province, Hubei Province, and Chongqing City mainly use traditional manual-transplanting paddy fields, including fish farming, snail farming, and crayfish farming in paddy fields. Jiangsu Province mainly uses advanced machine-transplanting techniques, including duck farming, fish farming, *Trionyx* farming, and crayfish farming in paddy fields. Heilongjiang and other northeastern provinces also use machine-transplanting techniques, including duck farming, fish farming, *Trionyx* farming, and crayfish farming in paddy fields. There are also loach farming, crab farming, and leech farming in paddy fields, but the area is not large, and the base farming still dominates. In terms of policy support, the No.1 Central Documents of 2016 to 2018 clearly expressed support for the development of integrated farming in paddy fields. Among them, the No.1 Central Document of 2018 stated that we should deeply promote the greening, premiumization, specialization, and branding of agriculture, and adjust and optimize the layout of agricultural productivity, which points out the direction for the further implementation of integrated farming in paddy fields. Meanwhile, the *National Agricultural Sustainable Growth Plan (2015–2030)*, the *National Agricultural Modernization Plan (2016–2020)*, and other planning guidance documents also put forward requirements for integrated farming in paddy fields, and the market prospect is very broad. At present, in Jiangsu Province, with the support of relevant policies and projects, the area of integrated farming in paddy fields has grown to 26,700 hm^2 in 2017.

4. Benefits of developing integrated farming in paddy fields

The implementation of integrated farming in paddy fields is an ecological circular agricultural measure that stabilizes grains, increases efficiency, and is safe and environmentally friendly. According to a survey, after the implementation of integrated farming in paddy fields, the use of chemical fertilizers and pesticides can be reduced by 20%–40% per mu. Integrated farming in paddy fields according to local conditions can not only improve the quality of rice, duck, fish, and other agricultural products, and reduce rural non-point source pollution, but also help to achieve high-quality agricultural products, save costs and increase benefits, and enhance the fun of tourism agriculture. It is a beneficial measure for the current township revitalization strategy to promote industrial enrichment of the people and increase the income of farmers.

II. Development Status and Trend of Integrated Farming in Machine-transplanting Paddy Fields

1. Distribution and status of integrated farming in machine-transplanting paddy fields

The integrated farming in machine-transplanting paddy fields is mainly distributed in Jiangsu Province, Zhejiang Province, Shanghai City, and Anhui Province in the Yangtze River Delta, as well as Heilongjiang, Liaoning, and Jilin Provinces in the Northeast, where machine-transplanting techniques are popularized. Among them, the farming modes in Jiangsu Province mainly include duck farming, fish farming, *Trionyx* farming, and crayfish farming in paddy fields. The farming area is mainly concentrated in northern Jiangsu, which has a relatively good aquaculture foundation. In recent years, such modes have also developed rapidly in southern Jiangsu, which have gradually expanded with the promotion of green agriculture. Heilongjiang and other northeastern provinces also use machine-transplanting techniques.

2. Future development trend

Since the market positioning of integrated farming in paddy fields aims at the mid-to-high end, it focuses on green, high-quality, safe, and delicious food, and achieves greater economic benefits in a smaller space, thus truly realizing green agriculture and industrial benefits. The development trend mainly shows the following features.

(1) Standardization and premiumization. Consumers increasingly recognize

safe, high-quality, and high-end agricultural products, and market demands are increasing, stimulating the development of relevant industries that produce high-quality agricultural products. To seize the opportunity, first, relevant government departments should strengthen the guidance, study, and formulate development plans with local characteristics of Jurong, Jiangsu, and introduce supporting policies; second, agricultural technique promotion departments should strengthen the introduction, test, and demonstration of new techniques and new varieties, apply more, better, and more environmentally friendly products and techniques to the integrated farming in paddy fields, adopt various teaching methods, enable farmers to learn and use new techniques well, and adress farmers' concerns about the future; third, relevant government departments should work with farmers to gradually formulate corresponding local standards, supervise the safe production of agriculture, ensure the high standardization and traceability of products in production, supply, processing, and sales, and achieve large-scale and standardized production. Promote the healthy development of the integrated farming industry in China.

(2) Industrialization. At present, integrated farming in paddy fields in most areas is still in the development stage, and the linkage between related planting and breeding industries is not strong. Next, the focus will be building and deepening the production, processing, operation, development, and other related industries in the local area, driving the development and increasing the planting benefits through the integration of the primary, secondary, and tertiary industries. For example, through the establishment of a rice industry consortium, it has initially realized the combination of production, supply, processing, and sales by leading small enterprises with big enterprises and supporting weak enterprises with strong enterprises. In industrial development, attention should be paid to the driving role of rural tourism, because paddy fields are not only farmlands, but also part of the beautiful ecological environment. By painting cartoon patterns on the duck house, the farmland landscape is beautified, and at the same time, the construction of tourist spots and rest stations converts the traditional single planting mode of paddy field to sightseeing and leisure. A large number of tourists are attracted to visit and play. They can also carry out various agricultural activities such as planting

rice seedlings, catching fish, catching ducks, and harvesting in the paddy fields. It attracted citizens to walk out of the ponderous reinforced concrete to the green fields full of life and experience the profound traditional culture and the unique rural tourism in these agricultural activities. Leisure and sightseeing extend the industrial chain, and make full use of the geographical advantages of being located in southern Jiangsu which has beautiful scenery of mountains and rivers. This promotes the shift of income source of agricultural production from sales of agricultural products to services and scenery, which increases the additional value.

(3) Branding. On the basis of producing high-quality agricultural products, we should strengthen publicity efforts, make good use of online and offline publicity channels, build regional brands, actively organize relevant enterprises to participate in various expos, promote the improvement of local rice brand awareness, and drive the benefits through the brand premium. With the advantage that integrated farming in paddy fields has been developed for several years, the product certification of "pollution-free, green, and organic agricultural products and PGI products" is encouraged. The development of "pollution-free, green, and organic agricultural products and PGI products" is continuously promoted towards "regional brands, product brands, and enterprise brands". The resources of different bases are integrated to develop corresponding products and sales strategies for different consumer groups. The whole production process, from rice breeding and transplanting to harvesting and sales, is market-oriented and meets the specific demands of consumers at multiple levels. The income source of agricultural production is expanded through brand development, which further strengthens local brands. Online and offline channels such as "Internet Agriculture" and "Rural Taobao" are used to increase the influence of integrated farming in paddy fields and improve the benefits for farmers.

Chapter II Integrated Farming Modes and Key Techniques in Paddy Fields

Section I Integrated Rice-Duck Farming Technique

Integrated rice-duck farming refers to an agricultural production mode in which ducklings (after receiving water acclimation) are stocked day and night in the transplanting fields with rice seedlings after establishment and in the early filling stage (Fig. 2-1), so that rice and ducks coexist and grow in the same ecological environment. Among them, the selected shelducks specifically refer to the duck species that replace manual intertillage, weeding, pest control, and other efforts in paddy fields.

Fig. 2-1 Integrated Rice-Duck Farming

I. Yield & Quality Indicators

(1) Yield indicator: 500–750 kg rice and 20–30 kg ducks per mu.

(2) Rice quality indicator and product safety standard: meet the standard NY 5115–2002. Quality standard: meet the standard for high-quality rice of grade III and above in GB/T 17891–2017.

(3) Safety indicators: inorganic arsenic ⩽ 0.15 mg/kg, lead ⩽ 0.2 mg/kg, cadmium ⩽ 0.2 mg/kg, fenitrothion ⩽ 1.0 mg/kg, triazophos ⩽ 0.05 mg/kg, dimethoate ⩽ 0.05 mg/kg, chlorpyrifos ⩽ 0.1 mg/kg, glyphosate ⩽ 0.1 mg/kg, butachlor ⩽ 0.5 mg/kg, dimehypo ⩽ 0.2 mg/kg, buprofezin ⩽ 0.3 mg/kg, phosphide ⩽ 0.05 mg/kg, and aflatoxin B1 ⩽ 0.01 mg/kg.

II. Environmental Conditions of Production Area

It should meet the requirements in *Green Food - Environmental Quality for Production Area* (NY/T 391–2013), and be certified by a qualified certification authority.

(1) Basic requirements for the ecological environment in the production area:

① The production area should be located in areas with a good ecological environment and without pollution and be away from industrial and mining areas, trunk highways, trunk railways, living areas, and pollution sources.

② The production area should be more than 50 m away from highways, railways, and living areas, and more than 1 km away from industrial and mining enterprises.

③ The production area should be away from the pollution sources, and measures should be taken to stop the entry of toxic and harmful substances into the production area.

④ The production area should be free from the threat of external pollution. No industrial and mining enterprises discharging toxic and harmful substances should be located in the upwind direction of the production area and upstream of the irrigation water. The irrigation water source should be deep well water, reservoirs, or other clean water sources, and polluted surface water such as sewage or pond water should not be used. The soil should not be improved by applying industrial waste residues containing toxic and harmful substances.

⑤ Biological habitats should be established to protect genetic diversity, species diversity, and ecosystem diversity and maintain the ecological balance.

⑥ It should be ensured that the production area has sustainable production capacity and will not cause pollution to the environment or other surrounding organisms.

⑦ The air quality of the production area is comprehensively analyzed using the air quality data of the production area in the previous year.

(2) Isolation zone: Between green food production areas and conventional production areas, isolation zones or buffer zones should be provided or distance should be kept to avoid contamination. ① Provide physical barriers; ② Use natural boundaries such as surface water and mountains; ③ Plant trees 5–10 m away from the edge of the green food planting area as double fences, with an isolation zone width of about 8 m.

(3) The quality of air, water, and soil should meet the general requirements of environmental quality in the production area: The daily average concentration of total suspended particles in the air is $\leqslant$ 0.30 mg/m^3; the daily average concentration of sulfur dioxide is $\leqslant$ 0.15 mg/m^3, and the 1-hour average concentration is $\leqslant$ 0.50 mg/m^3; the daily average concentration of nitrogen dioxide is $\leqslant$ 0.08 mg/m^3, and the 1-hour average concentration is $\leqslant$ 0.20 mg/m^3; the daily average concentration of fluoride is $\leqslant$ 7 μg/m^3, and the 1-hour average concentration is $\leqslant$ 20 μg/m^3.

For integrated farming, the water quality should meet the requirements of both farmland irrigation water and fishery water.

The soil quality should strictly follow the standard values, i.e. total cadmium $\leqslant$ 0.30 mg/kg, total mercury $\leqslant$ 0.25 mg/kg, total arsenic $\leqslant$ 15 mg/kg, total lead $\leqslant$ 50 mg/kg, total chromium $\leqslant$ 120 mg/kg, and total copper $\leqslant$ 50 mg/kg.

III. Cropping Pattern

Paddy fields are used for crop rotation of green manure and rape in winter.

IV. Variety Selection

1. Rice varieties

High-yield, high-quality, and multi-resistance rice varieties should be used, and their quality should meet the standard for high-quality rice of grade III and above in GB/T 17891–2017. High-quality rice is divided into high-quality indica rice and high-quality japonica rice. Among them, high-quality indica rice is further divided into long grain (> 6.5 mm), medium grain (5.6–6.5 mm), and short grain (< 5.6 mm) depending on the length of coarse rice. The grading indicators of high-quality rice include the head rice yield, chalkiness degree, and sensory evaluation score, and the limit indicators include the amylose content. For indica rice, the head rice yield should be more than 44% for long grains, 46% for medium grains, and 48% for short grains; the chalkiness degree should be less than 8.0%; the sensory evaluation score should be higher than 70, and the amylose content (dry basis) should be 14%–24%. For japonica rice, the head rice yield should be more than 55%; the chalkiness should be less than 6.0%; the sensory evaluation score should be higher than 70, and the amylose content (dry basis) should be 14%–20%.

2. Duck species

It should select shallow-feathered shelducks suitable for duck-rice farming with strong vitality, good stress resistance, long activity time, wide recipe, and omnivorous nature.

Shelduck, a species of the genus *Tadorna*, order Anseriformes, is distributed in inland freshwater areas of Eurasia, Africa, and Oceania. Shelducks are somewhere between ducks and geese. Both male and female shelducks usually have bright feathers and prefer living on land. Shelducks inhabit rivers, lakes, estuaries, ponds and nearby grasslands, wastelands, swamps, beaches, farmlands, and open forests on plains. They especially like lakes on plains and mainly live in inland freshwater. Sometimes, they are also found on beaches and saltwater lakes on the seaside and open grasslands far away from waters.

The common species of shelducks in China include Gaoyou shelduck, Shaoxing shelduck, Wuchuan shelduck, Weishan shelduck, Jinyun shelduck,

Youxian shelduck, and ruddy shelduck.

V. Seedling Culture and Duckling Breeding

1. Rice seedling culture

For seedbed selection, select soil with a clean water source, convenient for irrigation, and pollution-free as the seedbed.

The ratio of seedling bed to transplanting field, is 1 : 20 for dry rice seedlings, and 1 : 100 for machine-transplanted rice seedlings.

For seedbed fertilization or nutrient soil preparation of dry rice seedlings, 1,500–2,000 kg of straw scraps and 1,500–2,000 kg of livestock manure should be applied per mu before winter for seedbed fertilization, and plowed into the soil layer of 0–20 cm for 3 times. In spring, 1,500–2,000 kg of decomposed organic fertilizer should be applied per mu and plowed into the soil to mix evenly. Thirty days before sowing, 20–35 kg of urea, 65–100 kg of calcium superphosphate, and 25–35 kg of potassium chloride should be mixed and applied per mu by spraying on the bed surface in 3 applications, then fully harrowed to evenly mix the fertilizer into the soil layer of 0–10 cm. Preparation of nutrient soil: Before sowing, select fertile soil of the same soil type as the soil for seedbed fertilization or seedbed soil, sieve the soil with a sieve with a diameter of 5 mm, and prepare 6,500–10,000 kg fine soil per mu.

For machine-transplanting seedbed fertilization or nutrient soil preparation, seedbed fertilization can refer to the fertilization of watered rice seedlings. Preparation of nutrient soil: Select sandy loam that is basically free of weeds and pests, crush it and then sieve it with a sieve with a diameter of 5 mm, and prepare enough fine soil of 150 kg per mu of seedling bed. Thirty days before sowing, 350 kg of 45% compound fertilizer (N : P : K=15 : 15 : 15) should be added per 100 kg of fine soil, fully harrowed 2–3 times, and then composted for fertilization.

Seedbed specification of dry rice seedlings: the seedbed border is 1.2–1.4 m wide, the ditch is 0.2 m wide and 0.3 m deep; for machine-transplanted rice seedlings: the border surface is 1.4 m wide, the ditch is 0.25 m wide and 0.15 m deep.

Seed treatment: After being exposed to the sun and screened, every 3–4 kg of seeds are mixed with chemicals for *Gibberella fujikuroi* (Saw.) Wollenw and *Aphelenchoides besseyi* Christie and growth regulators, for example, 2 mL of 25% Sportak (prochloraz) + 2 mL of 10% Banzhongling (dithiocyano-methane) + 3 g of 15% paclobutrazol + 5 kg of water, and soaked for 48–72 hours. Then, take out the seeds for pregermination until white buds are shown before sowing.

Sowing period of dry rice seedlings: From May 20 to June 10; seedling age of machine-transplanted rice seedlings is 20 days.

Sowing rate per mu of seedling bed: 60 kg of dry rice seedlings or 210 kg of machine-transplanted rice seedlings.

Sowing process of machine transplanting: Spreading in trays (25 trays per mu of transplanting field) → spreading 2.0-cm-thick nutrient soil → sprinkling sufficient water → evenly sowing → covering nutrient soil of 0.5 cm → covering the mulching film → spreading grass on the film.

Seedling bed management: For machine-transplanted rice seedlings, the soft or hard tray seedling technique is used for seedling bed management. Green prevention and control of diseases, pests, and weeds in seedling beds should be carried out depending on the grass varieties, species of diseases and pests, and occurrence degree.

2. Breeding of ducklings

(1) Duckling preparation. The hatching time of breeding eggs is 28±1 days before the rice transplanting. The ducklings must be quarantined by the animal epidemic prevention and quarantine institution in the production area and should meet the requirements of GB 16549–1996.

(2) Ducklings health examination. Population examination items include: ① Static examination: mental status, appearance, nutrition, upright posture, respiration, ruminant state, feather, crown, and beard; ② Dynamic examination: the state of the head, neck, waist, and back, as well as limbs during movement; ③ Eating examination: the state of eating, chewing, and swallowing; ④ Defecation examination: the posture during defecation, as well as the quality, color, mixture, and odor of feces and urine. Individual examination should be made for individuals with abnormalities found in population examination. Regardless of whether abnormalities are found in population examination, 5%–20% of individ-

uals should be sampled for examination. Individual examination items include: ① Visual examination: mental appearance, nutritional condition, the posture of rising and lying movements, rumination, skin, fur, feathers, crown, beard, respiration, visual mucosa, natural apertures, rhinoscope, feces, urine, etc.; ② Palpation: the temperature and elasticity of skin, the sensitivity of thorax and abdomen, characteristics of crop contents, as well as the size, character, hardness, activity, and sensitivity of body surface lymph nodes (rectal examination should be made when necessary); ③ Percussion: the sound, position, and boundary of the heart, lung, stomach, intestine, and liver areas, and sensitivity of the chest and abdomen; ④ Auscultation: quack sounds, coughing sounds, heartbeats, and breath sounds of alveolar and tracheal, as well as sounds of gastrointestinal peristalsis; ⑤ Examination of the body temperature, pulse, and respiratory rate; ⑥ Examination of the color, quality, and odor of exudate, leakage, secretion, and pathological products.

(3) Centralized breeding. The ducklings are put into a breeding room at 28–30°C for centralized breeding, and the room temperature will be gradually reduced with the increase of breeding age until they adapt to the external temperature. It is necessary to keep ventilation, and feed water and food as soon as possible. The water should be prepared into 1‰ aqueous solution with 2% ciprofloxacin hydrochloride, and fed to the ducklings for 3 consecutive days. It should feed the ducklings with complete feed or commercial duck feed 4–6 times a day, and control the density below 50 ducklings per square meter, and the quantity per group should be within 150 ducklings. The padding in the breeding room should be replaced frequently to keep it clean and dry (Fig. 2-2).

(4) Water acclimation. Ducklings should receive water acclimation on sunny days, with a pool depth of 15–20 cm, and one side of the pool is an inclined plane of about 30°. The best time for the first water acclimation is about 10:00. After about half an hour of water acclimation, drive all the ducklings out of the water, and let them comb their feathers and rest in the sun. The second water acclimation should start at about 15:00 and be extended appropriately until the ducklings can swim freely in the water and their feathers become dry after leaving the water. For

ducklings with poor constitutions and whose feathers cannot dry after a long time, manually dry their feathers to reduce mortality.

Fig. 2-2 Ducklings

VI. Farming Management in the Transplanting Field

1. Field selection

The fields with flat blocks, which are convenient for irrigation and have good water and fertilizer conservation performance, should be selected as the production land for integrated rice-duck farming.

2. Preparation of transplanting field

Reinforcement of ridges: After the previous crops are harvested, the ridges around the integrated rice-duck farming area should be immediately heightened, widened, and reinforced by tamping to avoid water seepage and leakage. Generally, the ridges are required to be widened to 0.8–1 m and heightened to more than 0.3 m.

Apply sufficient basal fertilizer, and apply not more than 2,000 kg green manure per mu before dry tillage. The total chemical nitrogen, phosphorus, and potassium should be 14–15 kg of N, 7–8 kg of P_2O_5, and 7–8 kg of K_2O.

Fine soil preparation: 5–7 days before transplanting, conduct the dry tillage to sun the upturned soil, and then fill with water and rake the soil evenly. The machine-transplanting seedling fields need to be compacted for 1–2 days after raking.

3. Rice cultivation management

Transplant at an appropriate time: about 20 days for machine-transplanted rice seedlings.

The transplanting specification is 30 cm×12 cm, about 70,000–80,000 basic seedlings for machine-transplanted rice seedlings.

Fertilization: nitrogen fertilizer is applied by 30%, 40%, 20%, and 10%, respectively; all phosphate fertilizer is applied as basal fertilizer; 50% of potassium fertilizer is used as basal fertilizer, and the other 50% is used as flower promotion fertilizer.

Slurry management: After rice is transplanted and before ducks leave the field, a water layer of 5–10 cm should be maintained. The depth should increase with the growth of shelducks. After the ducks leave the field, intermittent wet irrigation and alternate wet and dry conditions should be maintained to raise old rice.

Disease and pest control: When the shelducks are in the field, they can provide biological control. After the ducks leave the field, biopesticides or pesticides that meet the technical specifications of DB32/T 343.2–1999 should be used for timely prevention and control according to the condition of rice diseases and pests.

For pest control, the plant protection policy of "prevention first, comprehensive control" should be implemented, and measures such as agricultural control, biological control, physical control, and chemical control should be comprehensively applied to control the occurrence and hazard of pests for the stability of the paddy field ecosystem. The use of pesticides should comply with GB 4285 and GB/T 8321. Inorganic arsenic, organic arsenic, organic tin, organic mercury, organic heterocyclics, fluorine preparations, organic chlorine, haloalkanes, organic phosphorus, carbamates, dimethylformimidamide, pyrethroids, substituted benzenes, diphenyl ethers, and sulfonylureas are prohibited. Commonly used pesticides for pollution-free rice production include trichlorfon, buprofezin, dimehypo, monosultap, triazoline, chlorpyrifos, fenobucarb, bacillus thuringiensis, carbendazim, tricyclazole, isoprothiolane, hymexazol, thiophanate-methyl, validamycin, triadimefon, bismerthiazol, thiediazole copper, antimicrobial agent

402, prochloraz, thiobencarb, butachlor, molinate, butyl·benzyl, oxadiazon, bentazone, dimepiperate, pretilachlor, quinclorac, bensulfuron methyl, pyrazosulfuron-ethyl, and glyphosate. These chemicals should be applied in limited dosages and at safe intervals. Rational mixing and alternate application of chemicals with different mechanisms of action or negative cross-resistance can overcome and delay the emergence and development of drug resistance to pests and diseases. The safe drainage period is 5–7 days.

4. Protection around fields

Piles and nets: Next to the field ridges around the integrated rice-duck farming area, piles made of bamboo tips or miscellaneous tree sticks are driven every 3 m. The piles are enclosed by the nylon net for protection, with a height of 1 m and a mesh of 2–3 cm^2. In addition, every 4–5 mu is an isolation square, in which piles and nets are used for isolation to control the activity range of the ducks.

Duck sheds: A simple duck shed of 4–5 m^2 should be built near the ridge of each isolated square as a place for shelducks to feed or rest, and avoid severe weather such as storms (Fig. 2-3).

Initial stocking area: A 3–5 m^2 initial stocking area should be enclosed with a nylon net in the paddy field in front of each simple duck shed so that ducklings can adapt to the environment.

Special protection: Special protection measures should be taken in areas with many natural enemies, such as weasels and muskrats.

Fig. 2-3 Simple Duck Shed

5. Stocking of shelducks

The stocking time is 7–10 days after the manual transplanting and about 15 days after the mechanical transplanting of rice seedlings. The working ducklings that have received water acclimation for 5–7 days should be stocked in time. The ducklings should be stocked in the initial stocking area for 2–3 days.

Number of stocking ducklings: 15–18 working ducklings are stocked per mu, and 60–90 ducklings are stocked in each isolation zone.

6. Feeding management of shelducks

Duckling stage: The shelducks should be fed 3 times a day in the morning, noon, and evening within 15 days after stocking in the paddy fields, and the feed should be supplemented appropriately. The feed should comply with NY 5032–2001. Medium duck stage: When the shelducks grow to about 0.5 kg, they should be fed twice in the morning and evening, and the feeding amount should be less in the morning and sufficient in the evening. The feed mainly includes wheat and blighted grain, with an appropriate amount of medium shelduck premix. Fattening stage: The shelducks are fattened in the field 15–20 days before rice heading. The feeding frequency is 2–3 times a day, mainly with adult shelduck pellet feed, which is mixed with some wheat, crushed rice, etc., and the feeding amount should be sufficient.

7. “Three preventions” of shelducks

(1) Disease prevention.

The prevention and control of duckling disease after hatching should be carried out under the guidance of local veterinary departments, and the application of veterinary drugs should meet the requirements of NY 5030–2001 guidelines. For duck disease prevention, prevention measures should be taken according to the prevalence of local diseases, and comply with NY 5263–2004 guidelines.

① Veterinary prescription drugs should be applied based on the prescriptions issued by veterinarians and in accordance with the relevant provisions of the Regulations on the Administration of Veterinary Drugs, and the prescription should be maintained for more than 3 years.

② Antibacterial drugs should be used carefully. Drug sensitivity tests should be carried out before medication, and narrow-spectrum antibacterial drugs are preferred over broad-spectrum antibacterial drugs. The results of drug sensitivity tests should be archived. Meanwhile, alternate medication should be considered to minimize drug resistance.

③ Drugs should be applied in strict accordance with the veterinary drug labels and package inserts approved by the Ministry of Agriculture and Rural Affairs, including the route of administration, dose, course of treatment, animal species, indications, withdrawal period, etc.

④ Veterinary drugs should not be used beyond the scope of their package inserts, drug varieties prohibited and not allowed by the Ministry of Agriculture and Rural Affairs should not be used, and human drugs should not be used; expired or deteriorated veterinary drugs should not be used, and bulk drugs should not be used.

⑤ Feed drug additives should be used according to the No. 168 Announcement of the Ministry of Agriculture of the People's Republic of China.

⑥ Adrenomimetics, antiasthmatics, anticholinergics, cholinomimetics, glucocorticoids, antipyretics, analgesics, and antiphlogistics should be used carefully and in strict accordance with the approved effects, purposes, usage, and dosage.

⑦ For non-clinical medical purposes, narcotics, analgesics, sedatives, central stimulants, sex hormones, chemical immobilizers, and skeletal muscular relaxants should not be used.

The use of veterinary drugs should be recorded in detail. The record should at least include animal species, year (day) age, weight, quantity, diagnostic results or purpose of the medication, drug name, specification, dose, route of administration, course of treatment, manufacturer, product approval No., production date, batch No., etc. Units or individuals using veterinary drugs should maintain drug use records for more than 3 years. In addition, the treatment of duck eggs produced during the withdrawal period should be recorded. The efficacy and adverse effects of veterinary drugs should be observed and recorded. When animals die, professional veterinarians should be invited for a necropsy to analyze whether it is caused by drugs or diseases.

Disease prevention should be carried out from three aspects: First, environmental sanitation conditions. The environmental sanitation quality should meet the requirements of NY/T 388–1999, and sewage and dirt treatment should meet the national environmental protection requirements; Site selection, building layout, facilities, and equipment should meet the requirements of NY/T 5264; The harmless treatment and disinfection of diseased carcasses should be carried out according to the requirements of GB 16548–2006 and GB/T 16569–1996 respectively. Second, breeding management. The introduced ducks should come from qualified breeding duck farms approved by the animal husbandry and veterinary administrative department, and animal quarantine certificates should be obtained. Vehicles and appliances used to transport ducks must be thoroughly cleaned and disinfected, and certificates of carrier disinfection for animals and animal products must be obtained. After ducks are introduced, they should be quarantined for 7–14 days, and the quarantine can only be ended after their health is confirmed. The feeding management, daily disinfection, and use of feed, veterinary drugs, and vaccines of shelducks should meet the requirements of NY/T 5264–2004, and regular supervision and inspection should be carried out. The drinking water for shelducks should meet the requirements of NY 5027–2008. Third, immunization. Shelduck farms should selectively carry out prophylactic vaccination of diseases according to the requirements of the *Animal Epidemic Prevention Law of the People's Republic of China* and its supporting regulations and in combination with the actual local epidemic situation. The selected vaccines should meet the requirements of the *Quality Standards for Veterinary Biological Products of the People's Republic of China*, and scientific immunization procedures and methods should be selected.

(2) Poisoning prevention.

The pesticide sewage from the surrounding non-integrated rice-duck farming areas should be strictly prevented from flowing into or permeating into the integrated rice-duck farming area. Once pesticide poisoning symptoms occur, detoxification treatment should be carried out in time. Dead ducks, dead fishes, and other carrion debris in the field should be removed in a timely manner to prevent botulinum toxin poisoning. Once duck death in batches is caused by botulinum toxin poisoning, ducks in the poisoning area should be driven to a safe environment for isolation and observation in time.

(3) Heatstroke prevention.

In the high-temperature season, the water layer in the fields should always be kept at about 10 cm to prevent heatstroke.

VII. Harvest

Harvest of ducks: 7–10 days after rice heading, shelducks should be withdrawn from paddy fields in time for processing and sale or be fattened in another place.

Harvest of rice: Rice should be harvested when the rice grains are yellow.

VIII. Expansion Model of Integrated Rice–Duck Farming

In order to make full use of the potential space-time and nutrient structure of the paddy field ecosystem and realize the efficient utilization of resources, production loop, gain loop, and product processing loop can be added on the basis of integrated rice-duck farming to improve the ecological benefits of paddy fields. At present, the main integrated farming modes include the seasonal ecological modes of "rice + duck + duckweed", "rice + duck + duckweed + fish", and the annual ecological mode of "rice + duck + grass + geese".

The ecological model of "rice + duck + duckweed" gives full play to the unique time-space advantages of the paddy field ecosystem and will introduce duckweeds with the advantages of not occupying space and not competing with rice for nutrients into the integrated rice-duck farming system. The duckweeds float on the water' s surface, fix nitrogen from the atmosphere, absorb the required nutrients from the water' s surface, and have the function of potassium enrichment. Placing duckweeds in paddy fields can meet 70% of the demand for the nitrogen and potassium nutrients of rice. Paddy fields provide a good growing environment for duckweeds and ducks, as the duckweeds not only provide sufficient feed for ducks but also return to the paddy fields through the stomach of ducks to improve the supply of soil nutrients in the paddy fields. The introduction of ducks also solves the problem of raising duckweeds in paddy fields in the past, controlling weeds and pests for rice and duckweeds, fertilizing them, and dividing and killing the duckweeds for fertilization. The ecological model of "rice + duck + duckweed" gives full play to the positive role of duckweed in the farmland ecosystem, intercepts the loss of matter and energy in the system, makes the ecological model of "rice + duck + duckweed" develop towards a virtuous cycle, and improves the ecological carrying capacity and material output capacity of the paddy field ecosystem.

The ecological model of "rice + duck + duckweed + fish" organically combines rice, ducks, duckweeds, and fish in paddy fields, and produces more pollution-free, high-quality agricultural products in the limited space-time structure of paddy fields. In this model, paddy fields provide sufficient growth and living space for ducks, duckweeds, and fishes. Weeds, pests, and duckweeds in the field can provide feed for ducks and fishes. Ducks and fishes can control weeds and pests for rice and duckweeds, fertilize them, kill the duckweeds for fertilization, and loosen soil for rice to stimulate growth.

The annual ecological model of "rice + duck + grass + geese" is to plant grass and breed geese after integrated rice-duck farming, thus building a year-round ecological farming model combining agriculture and animal husbandry. Field tests show that the forage grass-goose farming mode can effectively control winter weeds in the field and can replace herbicides. Compared with the rice-wheat farming model, the soil fertility and physicochemical properties are significantly improved. Moreover, forage grass is a nutrient-rich plant with high lysine, better protein quality than cereals, and rich mineral nutrition. Geese have a strong foraging ability, extremely developed digestive tract, and high digestibility, so they can make full use of forage grass.

Integrated rice-duck farming techniques have been applied to a certain extent in almost all rice regions across China. Many places have seen the technical, ecological, and food safety advantages of pollution-free integrated rice-duck farming, and have expanded the application area successively. However, some noteworthy problems have also arisen. On the one hand, rice production in China is carried out by households, and the measures taken by every farmer are not uniform when technology is applied. The measures are not in place, and the services of the agricultural technique promotion department are not in time. On the other hand, China's market economy is not yet mature, and high prices for high-quality products cannot be fully realized. In some places, it is difficult to increase the price of rice and ducks (including duck meat), especially since ducks are marketed in a centralized manner and are difficult to sell in a short time. In view of these conditions, the development mechanism of the integrated rice-duck farming technique in China should adopt the model of "government service (including subsidies) + farmer production (including self-production in enterprise bases) + company processing and sales". The government should issue policies to encourage farmers (including agricultural enterprises) to adopt the integrated rice-duck farming technique and coordinate relevant work. Enterprises should be responsible for the supply of pollution-free materials (including varieties) and the integration of technology. Farmers should organize rice production and duck breeding with the pollution-free integrated rice-duck farming technique according to the technical operation procedures of enterprises, and finally the enterprises will purchase rice and ducks at a higher price to ensure the increase of rice planting benefits for farmers. Through the comprehensive chain development integrating planting, breeding, processing, and sales, the brand advantage is created, so that the integrated rice-duck farming technique can be truly applied.

Integrated rice-duck farming inherits the essence of traditional Chinese agriculture, makes full use of and rationally protects natural resources, and effectively connects different agricultural industrial sectors. It is the best choice for sustainable rice production in the new century. China is a large rice-producing country and also a large duck–producing country. Integrated rice-duck farming has realized the strong combination and close integration of the two advantageous industrial sectors and has become an industry with the most advantages, potential, and vitality. China is also a big consumer of rice and ducks, with a huge consumer market for safe and high-quality rice and ducks. Modern rice production can only produce ordinary rice, while integrated rice-duck farming can produce organic rice, organic duck, and organic fertilizer at the same time. Modern rice production depends on the input of a large number of chemical fertilizers, pesticides, and insecticides, and it is difficult to be sustainable. Integrated rice-duck farming only needs the input of a small number of rice seeds, ducklings, and duckweed seeds to achieve the sustainable production of rice, ducks, and duckweeds. The input is greatly reduced, and the output is greatly improved.

Section II Integrated Rice-Crayfish Farming Technique

I. Integrated Rice-Crayfish Farming

Integrated rice–crayfish farming refers to the fact that in one year, the fields of integrated rice-crayfish farming can not only harvest one season of rice, but also cultivate one season of commercial crayfish. Crop clean refers to catching all crayfish each crop after they have matured or the bed of each crop of crayfish.

II. Environmental Conditions of Production Area

The environment of the production area should comply with the NY/T 5361–2016 and NY/T 5010–2016. Production areas free from industrial, agricultural, forestry, medical, and domestic waste and sewage, as well as other pollution, should be selected for integrated farming. Aquaculture water should be free of abnormal color, odor, and taste. Fields with sufficient water sources, convenient irrigation and drainage, good water quality, no pollution, and strong water conservation ability should be selected.

The basic indicators of irrigation water should include total mercury, total cadmium, total arsenic, total lead, and total chromium (hexavalent). In addition, according to the local environmental characteristics and the source characteristics of irrigation water, additional monitoring of the cyanide, chemical oxygen demand, volatile phenol, petroleum, total salt, and fecal coliforms can be performed.

Aquaculture water quality requirements: total coliforms ≤ 5,000/L, total mercury ≤ 0.0001 mg/L, total cadmium ≤ 0.005 mg/L, total lead ≤0.05 mg/L, total chromium (hexavalent) ≤ 0.05 mg/L, total arsenic ≤ 0.05 mg/L, petroleum ≤ 0.05 mg/L, volatile phenols ≤ 0.005 mg/L, sodium pentachlorophenate ≤ 0.01 mg/L, methyl parathion ≤ 0.0005 mg/L, dimethoate ≤ 0.1 mg/L , carbofuran ≤ 0.01 mg/L.

Soil environmental quality monitoring indicators are divided into basic indicators and optional indicators. Among them, the basic indicators include total mercury, total arsenic, total cadmium, total lead, and total chromium; the optional indicators include total copper, total nickel, and phthalates.

The environmental protection of freshwater aquaculture production areas will strengthen environmental conservation, implement protective measures, and prevent pollution. The signs of the production area are set up and clearly indicate the name of the production area, area, scope, and anti-pollution warnings.

III. Field Work Construction

1. Field selection

The paddy field should be selected in areas with flat terrain, sufficient water sources, no pollution sources around, separate and convenient irrigation and drainage, non-sandy soil, and firm and watertight ridges in the outermost area (Fig. 2-4). The environment and substrate should comply with the GB 15618–2018, NY/T 847–2004, and NY/T 5361–2016. The substrate should meet the requirements of total mercury $\leqslant$ 0.2 mg/kg, total cadmium $\leqslant$ 0.5 mg/kg, total lead $\leqslant$ 60 mg/kg, total chromium $\leqslant$ 80 mg/kg, total arsenic $\leqslant$ 20 mg/kg, and DDT $\leqslant$ 0.02 mg/kg.

Fig. 2-4 Rice-Crayfish Farming Fields in Jiangning District, Nanjing

2. Field works

The field for integrated rice-crayfish farming has a maximum area of no more than 60 mu, and the existing ridges and ditches should be used as much as possible.

The outermost ridges should be heightened to more than 60 cm above the field and widened to a width of 2 m.

3. Irrigation and drainage

The irrigation and drainage systems should be independent and should not be mixed. The water inlet should be located at the higher corner of the field, and the outlet should be at the lower corner. The water inlet is located on the ridge, and the water outlet is located at the lowest point at the bottom of the ring ditch and is sealed with an 80-mesh nylon net.

4. Escape prevention net

An escape prevention net should be provided on the outermost ridge, 20 cm above the ground, and made of plastic cloth, color steel tiles, etc. The net should be buried in the soil for 10–15 cm and compacted, leaving 20–30 cm on the ground, and wooden bamboo piles or stainless steel sticks should be erected every 2 m to secure and tension the net.

5. Field disinfection

Drug use for field disinfection should comply with the GB/T 19630.1–2011. Quicklime, bleaching powder, chlorine dioxide, tea seed cake, potassium permanganate, and microbial preparations can be used to disinfect aquaculture water and pond bottoms to prevent diseases. For example, after the transformation of the paddy field is completed, add water until the water surface is about 10 cm above the field surface, and apply 100 kg quicklime per mu to the water for disinfection.

IV. Planting Aquatic Plants

Elodea nuttallii (Fig. 2-5) should be planted on the field from November of the previous year to February of the next year. It is planted in strips, with a width of 8–10 m and a plant spacing of 5–8 m. Grasses are planted in shallow water areas. Dig holes first, put the grasses in, and then compact the soil; or directly fork the grasses into the soil. The coverage of aquatic plants should be 30%.

In the side ditch, aquatic plants such as *Elodea nuttallii, Hydrilla verticillata*, and *Alternanthera philoxeroides* are selected, and the planting area accounts for about 30% of the area of the side ditch. The planting time of *Elodea nuttallii* is given above. The planting time of *Hydrilla verticillata* should be in March, and the *Alternanthera philoxeroides* should be planted in spring when the water temperature is higher than 15℃.

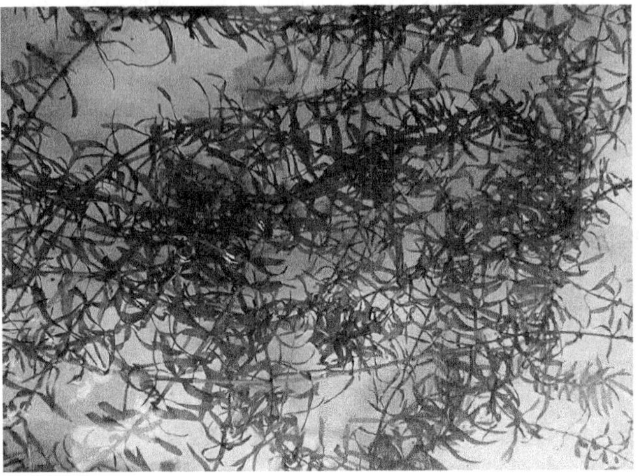

Fig. 2-5 *Elodea nuttallii*

V. Selection of Farming Varieties

1. Rice

It is advisable to select rice varieties with sturdy stems, strong tillering ability, lodging resistance, disease resistance, high-yield performance, excellent quality, and suitable for local planting. According to the rice adaptability of the Nanjing rice area, Nanjing 5055, Nanjing 3908, Nanjing 46, and other varieties with good quality and resistance should be selected.

2. Fingerling Crayfish

Healthy, high-quality fingerling crayfish (*Procambarus clarkii*) should be selected (Fig. 2-6). The fingerlings produced by the local enterprises with the license for the production and operation of aquatic fingerlings should be preferentially selected and pass the quarantine inspection. The transportation time of fingerlings should not exceed 2 hours.

The quality of fingerlings should meet the requirements. The specifications should be uniform; the best body color is bluish-brown, and light red comes second;

the appendages should be complete, and the body surface should be smooth; the reaction should be quick, and the activity ability should be strong.

Fig. 2-6 High-quality Fingerling Crayfishes

VI. Rice Cultivation

Rice cultivation management should comply with the relevant provisions of SC/T 1135–2017. The paddy field works should ensure the effective planting area of rice, protect the plowing layer of the paddy field, and the proportion of ditches and pits should not exceed 10%. There are about 18,000 holes per mu, with 4 or 5 rice seedlings per hole, and the row spacing and plant spacing are 30 cm × 12 cm.

For the standard planting of the integrated rice-fishery farming technique, the minimum target unit yield of rice should be set according to the technical indicator requirements. In the integrated farming mode, the marginal effect of rice cultivation should be exerted. Through the marginal dense planting, the maximum number of rice planting holes per unit area should be guaranteed. In the crop rotation mode, the stubble connection ensures the effective production cycle of rice and promotes stable rice yield.

1. Green fertilization of rice and aquatic plants

Return straw to the field: Leave stubbles of about 30 cm when the previous crop is harvested by machine. After harvesting, collect more than 1/2 of the crushed straws and pile them into a stack with a height of about 50 cm, so that all straws can be returned to the field and gradually decomposed. Return aquatic plants to the field: From May to June, after the crayfishes are harvested and marketed, the *Elodea nuttallii* is dehydrated for 2–3 days after being exposed to the sun, and then the rice machine can be directly used for transplanting, so that all the aquatic plants on the field surface are returned to the field. Apply organic fertilizer: Generally, 500 kg of decomposed organic fertilizer or about 100 kg of cake fertilizer should be applied per mu before rice machine transplanting. Basal application of chemical fertilizer: 30 kg/mu of 45% slow-release compound fertilizer is used as basal fertilizer. Moderate top dressing: 10 kg/mu in the rice tillering stage, 10 kg of compound fertilizer, and 5 kg of urea to promote and protect the spike. In the critical growth period of aquatic plants, topdress liquid bio-organic fertilizer of 20 kg/mu or urea of 5 kg/mu.

2. Green prevention and control of rice

Agricultural control: Choose high-quality, high-yield, disease-resistant, insect-resistant, and lodging-resistant rice varieties. Physical control: Install 1 solar insect killer lamp every 30 mu. Ecological control: Use pheromone traps to trap or interfere with the mating of pests (Fig. 2-7). One pheromone trap is installed every 200 m^2, which is set in a row. The pheromone core is replaced every 15 days, and the traps are 15–20 cm above the rice canopy. Input *Trichogramma* to control pests, and plant vetiver grass in the surrounding ridges to trap and kill pests, etc.

Fig. 2-7 Insect Killer Lamps and Pheromone Traps

VII. Farming of Crayfish

1. Input time and specifications of fingerling crayfish

Fingerling crayfish are generally input in mid-March and early April before rice transplanting; after the rice seedlings turn green, the second batch of fingerling crayfish can be added according to the remaining fingerling crayfish in the paddy fields. The specification of fingerling crayfishes is 200–300/kg.

2. Density of fingerling crayfish

The density of fingerling crayfish for crayfish farming is 4,000/mu. When inputting the first batch of fingerling crayfish, the number of fingerling crayfish with a specification of 3–4 cm should be 6,000–8,000/mu; the number of fingerling crayfish with a specification of 4–5 cm should be 5,000–6,000/mu. When inputting the second batch of fingerling crayfish, the number of fingerling crayfish with a specification of about 5 cm should be 2,000–4,000/mu.

3. Adult crayfish fishing

The general breeding cycle is 30–50 days, and the fishing and marketing are concentrated in mid- and late May (Fig. 2-8). The yield of crayfish per mu is generally 80 kg. After the fishing is completed, clear the pond for the first time.

Fig. 2-8 Farmers Fishing Crayfishes

VIII. Input Method of Fingerling Crayfish

Put fingerling crayfish on the abundant aquatic plants in the shallow water area, and let fingerling crayfishes enter the water by themselves.

Fingerling crayfish are generally transported in a plastic frame paved with wet aquatic plants. If the transportation time is short, the crayfish should be placed in shallow water or a place with more aquatic plants, so that they can enter the water by themselves. If the transportation time is long, the crayfish should be soaked in the paddy field water for about 1 min in advance, then lift them up and leave them for 2–3 min, and repeat this 2–3 times, so that the surface and gill cavity of the crayfish can absorb enough water, and then gently place the crayfish in shallow water or places with many aquatic plants, so that they can enter the water by themselves.

1. Feeding management

Feed requirements: The feed during the aquaculture of crayfish should comply with the GB/T 19630.1–2011.

Feeding method and amount: The feeding method is "fixed time, fixed place, fixed quality, and fixed amount; depending on the season, depending on the weather, depending on the water quality, and depending on the activity of crayfish". Start feeding a small amount on the third day after placing the fingerlings, and gradually increase the feeding amount as the growth accelerates. The feed should be basically eaten up within 2–3 hours after feeding. Under normal weather conditions, crayfish are generally fed twice a day in the morning and evening. Generally, about 30% of the daily feeding amount is fed before sunrise, and about 70% of the daily feeding amount is fed after sunset. The daily feeding amount is 3%–8% of the weight of the crayfish in the field. During rainy days, molting periods, disease periods, or overwinter periods, the feeding amount should be reduced or non–existend.

The types of feeds include vegetable feeds, animal feeds, and special compound feeds for crayfish. It is recommended to use the special compound feed for crayfish, and the compound feed should meet the requirements of GB

13078–2017 and NY 5072–2002. The feed should meet the requirements of feed hygiene indicators and be subject to tests including inorganic contaminants (total arsenic, lead, mercury, cadmium, hexavalent chromium, fluorine, nitrite), mycotoxins (aflatoxin B1, ochratoxin A, zearalenone, deoxynivalenol, T-2 toxin, fumonisins (B1+B2)), natural plant toxins (cyanide, free gossypol, isothiocyanate, oxazolidine thione), organochlorine contaminants (PCB, benzene hexachloride, DDT, hexachlorobenzene), microbial contaminants (total mold, total bacteria, salmonella) .

The raw materials used in the processing of fish feed should meet the requirements of various raw material standards, and raw materials that are damp, moldy, insect-infested, rotten, or contaminated by petroleum, pesticides, and harmful metals should not be used. Leather powder should be dechromized and detoxified. Soybean raw materials should be processed by inhibitors to destroy protease.

Feeds for organic aquaculture should be organic, wild, or licensed by a certification body. When the quantity or quality of organic or wild feed cannot meet the demand, conventional feed can be fed up to a maximum of 5% of the total feed. In the event of unforeseen circumstances, a maximum of 20% of the conventional feed can be fed in the year after the assessment and approval of the certification body. At least 50% of the animal protein in the feed should be derived from by-products of food processing or other products unfit for human consumption. In the case of unforeseen circumstances, this proportion can be reduced to 30% in the year. Natural mineral additives, vitamins and trace elements can be used; when the nutritional demands of aquatic animals cannot be met, the minerals, trace elements, and synthetical vitamins can be used. Substances that should not be added to feed or fed to aquatic organisms in any way include synthetic growth promoters, synthetic phagostimulants, synthetic antioxidants and preservatives, synthetic pigments, non-protein nitrogen, organisms of the same family as the farming object and their products, chemically purified amino acids, and genetically modified organisms or their products.

2. Disease prevention and control

The disease prevention and control in the farming process of crayfish should comply with the GB/T 19630.1–2011.

Preventive measures should be taken to ensure the health of farming objects. All management measures should be aimed at increasing the disease resistance of organisms. The farming density should not affect the health of aquatic organisms and should not cause abnormal behavior. The density of organisms should be monitored regularly and adjusted when necessary. Quicklime, bleaching powder, chlorine dioxide, tea seed cake, potassium permanganate, and microbial preparations can be used to disinfect aquaculture water and pond bottoms to prevent disease of aquatic organisms. Aquatic animal diseases can be prevented and treated using natural medicines. Conventional fishing medicines can be used on aquatic organisms in cases where preventive measures and natural medicine treatments are ineffective. Sick organisms should be isolated during routine medical treatment. Antibiotics, chemosynthetic drugs, and hormones should not be used for routine disease prevention of aquatic organisms.

3. Daily management

Patrol the fields every morning and evening to observe the changes in water quality of the paddy fields, as well as the eating, molting growth, activity, and the presence of diseases of the crayfish, and adjust the feeding amount in time. If a large number of crayfish are floating, climbing grass or slope, or have other abnormalities, prevention and control measures such as oxygenation facilities, water changes, and medication should be taken in a timely manner. Regularly inspect and maintain anti-escape facilities and deal with problems in a timely manner.

4. Water level control

Generally, the water depth of the border surface is kept at 30–60 cm. In the early stage of fingerling placement, shallow water, and temperature increase should be adopted, and the water level should gradually increase as the temperature rises. The field water layer should be 30–40 cm when fishing crayfish, as shown in Table 2-1.

Table 2-1 Water Level Control in Integrated Rice-Crayfish Farming

Period	Water level
January–February	About 50 cm above the field surface
March	About 30 cm above the field surface
April	About 40 cm above the field surface
May–Before the field preparation	About 50 cm above the field surface
Field preparation–July	About 5 cm above the field surface
July–September	About 20 cm above the field surface
Field drying period	About 30 cm below the field surface
7 days before rice harvest–Rice harvest	20–30 cm below the field surface
After rice harvest–November	About 30 cm above the field surface
November–December	40–50 cm above the field surface

5. Water quality regulation

During the fingerling breeding period, it is advisable to apply fermented and decomposed organic fertilizer, and the application amount is 100–150 kg/mu. Combined with the measures of supplementing fertilizer, adding water, and changing water, the transparency of the water body should be controlled at 25–35 cm during the whole farming period. At other times, according to the color of the water, the weather, and the activity of crayfish, the water quality can be adjusted by supplementing fertilizer, adding water, and changing water, so that the transparency of the water body is controlled at 35–45 cm.

IX. Fishing of Crayfish and Rice Harvesting

1. Fishing of crayfish

Ground cages and crayfish cages are fixed on the ring ditch or border surface in the evening and withdrawn in the morning. Every crop of crayfish should be caught and cleared as much as possible to be "crop clean".

When inputting fingerling crayfish, the first batch of adult crayfish is caught from late April to early June; the second batch of adult crayfish is caught from early August to the end of September. When introducing brooder crayfish, the fingerling crayfish are caught from mid-March to mid-April; the adult crayfish are caught at the same time as above when inputting fingerling crayfish.

The main fishing tools are ground cages. The mesh size of ground cages for fishing the fingerling crayfish is preferably 1.6 cm; the mesh size of ground cages for fishing the adult crayfish is preferably 2.5–3.0 cm.

In the early stage of fishing, place the ground cages on the field surface and in the side ditches, and change the place every 3–5 days. When the catch gradually decreases, lower the water level of the paddy field so that the crayfishes fall into the side ditch, and then place ground cages concentratedly in the side ditches. In the paddy fields used to breed fingerling crayfish, when the adult crayfish are caught in autumn, the fishing will be stopped when the daily catch is less than 0.5 kg/mu, and the remaining crayfish should be used to breed brooder crayfish.

2. Rice harvest

Judge the maturity stage performance of different rice, and harvest it in time. The harvest time is generally from mid- to late October to early November. Leave about 40 cm of stubble, collect the straw scattered on the field, and pile it into small haystacks.

X. Production Records

The integrated rice-crayfish farming should be recorded and archived throughout the process, and the records and files should be kept for more than 2 years, including the variety of sources of crayfish and rice, the sources and use of inputs, the process of rice-crayfish farming, the sales of crayfish and rice, etc.

Section III Integrated Rice-Crab Farming Technique

Rice-crab farming is based on the theory of symbiosis between rice and Chinese mitten crabs (Fig. 2-9). In the environment for rice-crab farming, crabs can remove weeds in the fields and eat pests, and their excrement can fertilize the fields and promote the growth of rice. At the same time, the rice provides abundant natural feed and good habitat conditions for the growth of river crabs, which are mutually beneficial and form a virtuous ecological cycle.

Fig. 2-9 Chinese Mitten Crab

I. Fingerling Crabs

The fingerling crabs are bred into young crabs with immature gonads of about 100–200/kg.

II. Product Indicators

The yield of rice is above 6,000–7,500 kg/hm^2, and the rice meets the green food standard of NY/T 419–2014. Commercial crabs should comply with the provisions of GB/T 19783–2005. The yield of commercial crabs should be 50–100 kg/mu, with an average size of more than 100 g/crab, free of diseases and residual limbs, and meet the green food standard of NY/T 841–2021.

III. Environmental Conditions

1. Environment of production area

The environment of the production area should comply with the requirements of NY/T 391–2013.

2. Quality of water source

The quality of the water source should conform to the provisions of GB 11607–1989.

3. Selection of paddy fields for crab breeding

The suitable unit breeding area of the paddy field is 2.5–4.0 hm^2, with a sufficient water source, convenient drainage and irrigation, and satisfactory water conservation performance.

IV. Field Works

1. Reinforcement for field ridge

The field ridge requires reinforcement and compaction, with a top height of 50–60 cm, a top width of 50–60 cm, and a bottom width of 100–120 cm.

2. Excavation of crab ditch

Crab ditches account for less than 10% of the total area of paddy fields. Circular ditches are excavated around the crab ditches, with a width of 2–3 m and a depth of 0.8–1.0 m. Crab ditches are appropriately added in the middle of the paddy field with a large unit area.

3. Anti-escape facilities

After the rice transplanting in the paddy field and before the stocking of fry crabs, an anti-escape wall will be set up with film, the lower end will be buried in the soil for 15–20 cm, and the aboveground part will be 60 cm high, and the water inlet and outlet of the paddy field will be installed with an anti-escape net.

4. Paddy field disinfection

Quicklime or bleaching powder will be used per hectare for the disinfection of field parcels. The application amount of quicklime is 100–150 g/m^2, and that of bleaching powder is 7.5–15 g/m^2.

V. Rice Cultivation

1. Variety selection

Suitable varieties with high yield, high quality, and multi-resistance should be selected.

2. Transplanting method

After the leveling of field parcels, mechanical rice transplanting should be carried out in due time, with a row spacing of 25 cm or 30 cm, a plant spacing of 12–14 cm for conventional rice, and 4–5 seedlings in each hole. The plant spacing for hybrid rice should be 18–20 cm, 1–2 seedlings in each hole.

3. Fertilization management

Fertilization for paddy fields should meet the requirements of NY/T 394–2021.

① Principle of sustainable development: The fertilizers used in green food production should not cause adverse effects on the environment and should be beneficial to the protection of the ecological environment and the maintenance or improvement of soil fertility and soil biological activity.

② Principle of safety and high-quality: Safe and high-quality fertilizer products should be used in green food production to produce safe and high-quality green food. The application of fertilizers should not cause any adverse consequences for crops.

③ Principle of chemical fertilizer reduction and control: On the basis of ensuring the effective supply of plant nutrients, the application amount of chemical fertilizers will be reduced, and the ratio of elements should be balanced. The application amount of inorganic nitrogen should not be higher than half of the demand of the current crop.

④ Principle of giving priority to organic fertilizer: In the process of green food production, the types of fertilizers should be mainly farmyard manure, organic fertilizers, and microbial fertilizers, supplemented by chemical fertilizers.

Fertilizers added with rare-earth elements, fertilizers with unclear ingredients and containing potential safety hazards, and unfermented and decomposed human and animal manure should not be applied.

Domestic waste, sludge, industrial waste containing harmful substances, fertilizers produced with genetically modified varieties and their by-products as raw materials, and fertilizers that are prohibited by national laws and regulations should not be used.

The application proportion of N, P, and K is 1 : 0.5 : 0.5, the total nitrogen is 210–240 kg/hm^2, the nitrogen fertilizer used as basal fertilizer is 30%, nitrogen fertilizer for tillering is 40%, and nitrogen fertilizer used as panicle fertilizer is 30%. All phosphate fertilizer will be used as basal fertilizer, half potassium fertilizer as basal fertilizer and the other half as spikelet-promoting fertilizer. During top dressing, crab ditches should be avoided and fertilizer should be applied in turns in different areas.

4. Water level management

The water level during the tillering stage will be controlled at 3–5 cm, and the water level during the booting stage will be kept at 10–20 cm. The water will be drained 10 days before harvesting. The water level will be deepened as far as possible without affecting the growth of rice during the whole paddy field breeding cycle, and field draining will not be carried out.

5. Prevention and control of diseases, pests, and weeds

Comply with requirements of NY/T 393–2021 and SC/T 1135.1–2017.

Principles to be followed for the prevention and control of harmful organisms in green food production include the following aspects.

① Based on maintaining and optimizing agroecological systems. Establishing environmental conditions that are conducive to the reproduction of various natural enemies and unfavorable to the occurrence of diseases, pests, and weeds, improving biodiversity, and maintaining the balance of agroecological systems.

② Priority should be given to agricultural measures. For example, selecting varieties resistant to diseases and pests, carrying out quarantine for seeds and seedlings, cultivating strong seedlings, strengthening cultivation management, intertillage and weeding, plowing, sunning of the upturned soil, fields cleaning, crop rotation of rice stubble, intercropping, and interplanting.

③ Physical and biological measures should be used as much as possible.

For example, soaking seeds in warm water to control seed-borne diseases and pests, capturing pests mechanically, mechanical or manual weeding, trapping and killing pests with lights, colorful sticky boards, sex attractants and food, releasing natural enemies of pests and pest control by raising ducks in paddy fields.

④ Rational use of low-risk pesticides when necessary. If there are no sufficiently effective agricultural, physical, and biological measures, apply pesticides according to relevant regulations on the premise of ensuring the safety of people, products, and the environment.

6. Harvesting

After the full maturity of rice grains, the river crabs will be caught first, and then the rice will be harvested.

VI. Breeding of River Crab

1. Selection of fingerling crab

Young crabs with neat size, strong vitality, intact body parts, disease-free, and shiny body colors will be selected.

2. Temporary breeding of fingerling crab

The temporary breeding pond: It should be built based on 10%–20% of the crab breeding area in paddy fields and ditches close to the water source, with a depth of more than 1 m and a water depth of 0.5 m.

Time of temporary breeding: From February to March, the purchased juvenile crabs should be temporarily cultured in the surrounding circular ditches, and the fingerling crabs should be put into paddy fields after the rice seedlings transplanted by machine return to green.

Put fingerling crabs into the pond: The temporary density is 4.5×10^4 crab/hm^2, and soak the crabs in 3%–5% NaCl solution or 20 mg/L potassium permanganate solution for disinfection for 5–8 min.

Temporary breeding management: Feeding high-quality bait, with complete formula feed as the main feed and animal feed as a supplement. The crabs are fed once a day, and the feeding amount accounts for 3%–5% of the total weight of the river crab. The feeding amount should be appropriately adjusted according to the consumption of river crabs; half of the pond water should be replaced for 7–10 days, and the water quality should be adjusted.

3. Cultivation of aquatic plants

Before the stocking of fingerling crabs, *Vallisneria spiralis*, *Alternanthera philoxeroides*, and other varieties will be transplanted in the ditches, with a coverage degree of about 30% by aquatic plants.

4. Stocking of fingerling crabs in paddy fields

The stocking density of fingerling crabs is 7,500–12,000 crab/hm^2.

5. Feeding management

The quality of feed should comply with provisions of GB 13078–2017 and NY/T 471–2023. The crude protein content of formula feed should be between 30% and 40% and the feed for crabs should be fresh and free of spoilage. Crabs are fed once a day in the evening, and appropriate adjustments can be made according to consumption and weather conditions. Ecdysone can be added 2–3 days before the molting period.

Feeds and feed additives should conform to the provisions of the quality standards for single feeds, feed additives, formula feeds, concentrated feeds, and premixed products of additives. Feed additives and additive premixed feed should come from enterprises with production licenses and be provided with product standards and reference numbers. From the sensory perspective, it is required that feeds should have the proper color, smell, and morphological characteristics, with uniform texture and free of mildew, spoilage, caking, worm-eaten, peculiar smell, and foreign matter. Formula feed should be nutritious and balanced among nutrients.

Feed raw materials can be green food that has passed the certification or products derived from the standardized production base of green food products, or products that have been certified by the green food institution, produced in accordance with the green food production mode and produced in a self-constructed base that meets the green food standard.

Feed materials produced with genetically modified methods should not be used. Follow the principle of not using feeds of homologous animal origin. Industrial synthetic grease is prohibited.

VII. Daily Management

1. Water quality regulation and control

The water should be changed once every 7–10 days, depending on the water quality in the field, and the water change should not exceed 10% each time. The time for water change should be controlled within 3 hours, and the difference in water temperature should not exceed 3°C. Before and after the molting, the water needs to be replaced frequently. During the peak stage of molting, the water can be injected appropriately without changing the water.

2. Frequent patrol of fields

Pay attention to the change in water quality, the growth of crabs, whether the food consumption is normal, whether there are dead crabs, and whether there is water leakage in field ridges; pay attention to preventing water snakes, rats, frogs, large birds, wild and miscellaneous fishes, and other enemy organisms.

3. Prevention and control of diseases

The prevention of crab disease needs to be given priority, and the disinfection process for fingerling crabs, bottom soils, and water quality requires strict control. The use of drugs should comply with the regulations of NY/T 755–2022.

VIII. Crab Fishing

Crab fishing begins in late September to keep the appendages of river crabs intact.

Section IV Integrated Rice-Loach Farming Technique

In addition to integrated farming techniques of rice-duck, rice-crayfish, and rice-crab, those of rice-fish, rice-loach, and rice-*Trionyx* are also in the development and experimental promotion. Similar to the aforementioned techniques, these integrated farming techniques also need to be carried out comprehensively from two aspects: rice cultivation and breeding. From the perspective of rice cultivation, suitable water management in each stage of rice growth is the basis of high quality and high yield. From the perspective of breeding, it is of great importance to choose the periods for water entering and fishing of fish, loach, and *Trionyx sinensis*, so as to ensure both the sound survival of fry and fingerlings of these breeding animals and their sufficient growth time.

I. Field Works for Rice-Loach Farming

For plain areas, 10–15 mu is recommended; for mountainous areas and hills, 1–5 mu is recommended.

Side ditches should be excavated at 1 m from the inner side of the field ridge according to the terrain and size of the field parcel and local conditions. The side ditches can be excavated into rectangular-shaped, U-shaped, L-shaped, I-shaped, and other shapes, with a ditch width of 1–2 m wide and a ditch depth of 0.5–1 m and a slope ratio of 1 : 1; a mechanical operation channel with a width of about 4 m should be reserved on the side with convenient traffic conditions. According to the size of the field parcel, cross-shaped or well-shaped field ditches should be excavated in the center of paddy fields, with a width of 30–40 cm and a depth of 30–40 cm, which should be connected with the side ditches. In addition, it is necessary to excavate the temporary breeding pit at the intersection of the water intake and the side ditch, accounting for 0.5%–1% of the paddy field area, with a suitable length-width ratio of 3 : 2 and a depth of 1–1.5 m. A layer of plastic film with a thickness of 0.1–0.2 mm should be laid at the bottom of the temporary breeding pit, and then a layer of soil with a thickness of 10–15 cm should be flatly compacted on the plastic film; a sunshade net should be provided above the temporary breeding pit, and the area of the sunshade net should reach 80% of the area of the temporary breeding pit.

The field ridges need to be heightened, widened, and reinforced. The field ridges should be 60–80 cm higher than the field surface, with a bottom width of

120 cm and a top width of 80 cm. Field ridges should be compacted without water leakage.

The water intake and outlet facilities should be relatively independent. The water intake should be built on the field ridge, 50 cm above the field surface; the water outlet should be built at the lowest position of the side ditch; the water intake and outlet of the paddy field should be arranged diagonally, and the water intake and outlet should be equipped with double-layer anti-escape nets. A mesh bag with a length of 1.5 m and a diameter of 0.3 mm (50 mesh) is recommended for the water intake and a polyethylene net with a mesh size of 0.4 mm (40 mesh) is recommended for the outer latake of the water outlet and the inner layer should be made into a fish fence with wire mesh with a mesh size of 0.4 mm (40 mesh).

Anti-escape facilities should be embedded on the inner side around the field ridge, and polyethylene mesh with a mesh size of 0.4–0.6 mm (30–40 mesh) should be adopted and should be 20–30 cm higher than the field ridge and water intake, fixed with wooden poles, small bamboo poles or other materials, and buried 40–50 cm under the soil, with four corners in an arc shape.

II. Selection of Fingerling Loach

Loach (*Misgurnus anguillicaudatus*): The fingerling loaches used in rice-loach farming should come from enterprises with the production and operation license of aquatic fingerlings and should be qualified for quarantine. The loach broodstock should be caught from natural waters without artificial stockings, such as rivers, reservoirs, and lakes, or obtained from the fry and fingerlings collected from natural waters after artificial cultivation or provided by original (improved) fingerling farms at or above the provincial level. The fry should be the artificially-reproduced fry of the above-mentioned broodstock, and should have the qualities of normal body color, uniformity in size, strong motility, and sensitive response to changes in external environmental conditions, and the malformation rate should be less than 1%. The disability rate should be less than 1%.

III. Common Diseases of Loaches and Diagnostic Methods

The common diseases of loaches and diagnostic methods are shown in Table 2-2.

Table 2-2 Common Diseases of Loaches and Diagnostic Methods

Disease name	Pathogen	Symptoms	Epidemic season	Diagnosis
Saprolegniasis	*Saprolegnia* spp. or *Achlya* spp.	White spots appear at the lesion position during the initial stage, and in severe cases, the hyphae stretch into cotton and flocculence-shaped growths	It is mainly prevalent in winter and spring and can occur in various sizes of loach	A preliminary diagnosis can be made based on symptoms
Gill rot disease	*Cytophaga columnaris*	The diseased fish swims alone and slowly, with poor appetite, black body color, rotten, and white gill filaments, exposed cartilage at the tip, and sludge and mucus on the gill. In severe cases, the inner epidermis of the operculum can be corroded, forming a transparent "small window"	This disease can occur in both fingerlings and commercial fish and is prevalent from April to October and is most prevalent in summer every year	A preliminary diagnosis can be made based on symptoms
Red fin disease	Bacteria	The epidermis near the dorsal fin falls off and is grayish-white, the anus turns red and the muscles begin to rot. In severe cases, the fin rays fall off and the diseased fish do not eat, becoming weak until death	From June to September every year, with the highest disease rate in summer	A preliminary diagnosis can be made based on symptoms
Enteritis	*Aeromonas punctata*	The abdomen of the diseased fish is enlarged and the anus is red and swollen. If the abdomen is slightly pressed, yellow mucus will flow from the anus. The diseased fish swims alone, acts slowly, and loses its appetite	The epidemic season for commercial fish is from June to September. This disease mainly occurs from May to June among fry and fingerlings	A preliminary diagnosis can be made based on symptoms

continued

Disease name	Pathogen	Symptoms	Epidemic season	Diagnosis
Trichodiniasis and chilodonelliasis	*Trichodina*, *Trichodinella*, and *Chilodonella cyprini*	In the early stage, the food intake is reduced, and in severe cases, the worms are densely covered and the fish is sluggish	The epidemic season is from May to August, mainly endangering fish fry and fingerlings	Use a microscope to examine the fish's body surface and mucus on gill filaments, *Trichodina* and *Trichodinella* can be seen under the microscope
Ichthyophthiriasis	*Ichthyophthirius multifiliis*	There are white dots distributed on the skin of the chest, back, caudal fin and body surface of the diseased fish When the fish becomes seriously ill, the body surface seems to be covered with a white film, and the fish swims slowly, loses its appetite and loses weight	This disease mostly occurs in early winter, late spring, and late autumn, with water temperatures of 15–25°C, mainly endangering fish fingerlings	Use a microscope to examine the fish's body surfacc and mucus on gill filaments, and white dots are visible to the naked eye when the fish is seriously ill

IV. Transportation and Stocking of Fingerling Loach

Before transportation, fingerling loaches should be disinfected with 3%–4% mixture of salt and sodium bicarbonate (1 : 1) and soaked for 5–10 min. Oxygen bags should be used for oxygenating and transportation, with each bag containing 3 kg of water and 800–1,000 fingerling loaches of 3–4 cm. The transportation time should not exceed 6 hours.

After seedlings are transplanted for 10–20 days, fingerling loaches seeds will be stocked. The fingerling loaches for stocking should have a strong physique, smooth body surface, disease-free, injury-free, and strong mobility. When releasing fingerling loaches into the water, the temperature difference should be adjusted. The oxygen bag can be placed into the water body to be put in for about 30 min to make the temperature difference between the inside and outside of the oxygen bag ≤ 2°C, and then open the oxygen bag to let the fingerling loaches swim into the water body. In the central and southern areas of China, it is suitable to stock 10,000–15,000 fingerling loach/mu with a size of 3–4 cm/fingerling loach, and in the northern areas, it is suitable to stock 5,000–8,000 fingerling loach/mu with a size of 7–8 cm/fingerling loach.

Around 10 days before the stocking of fingerling loaches, fermented organic fertilizer of 200–250 kg/mu should be applied according to the area of ditches and pits and should be mainly applied within the ditches and pits. Powder feed should be fed for 10–15 days, crushed feed should be fed for 20–30 days, and then changed to pellet feed for continuous feeding, and the protein content should be 25%–30%. Using the bait resources in paddy fields to reduce the feeding amount, the daily feeding amount should be 1%–3% of the weight of the loach. The feeding amount needs to be reduced on cloudy days and low air pressure weather. The amount of feed fed each time should be consumed within 1–2 hours. Do not feed when the water temperature is higher than 30°C or lower than 20°C. The feeding location should be selected in the side ditch and temporary breeding pit, and the feeding should be carried out once at 9:00 and 17:00 every day. Feeding should be regular, positioned, qualitative, and quantitative.

Section V　Restrictions and Countermeasures on the Development of Integrated Rice-Fishery Farming Technique

I. Water Conservancy Conditions

Paddy fields with sufficient water sources and convenient irrigation and drainage are the primary conditions for developing fish farming in paddy fields. Due to geographical reasons, water sources are generally scarce, the land is relatively dry, soil leakage is common, and water conservation performance is poor in North China, which has caused great difficulties for the development of fish farming in paddy fields there; South China is susceptible to flood disasters due to rainy weather, which brings unexpected losses to fish farming in paddy fields.

II. Production Technique Level Needs to be Further Improved

Although China has a long history of fish farming in paddy fields, its technical foundation is weak. In recent years, many new experiences have been created in the field of fish farming in paddy fields, generating a mode of planting and farming plus three-dimensional ecological agriculture, and the techniques of fish farming in paddy fields have become a comprehensive technical discipline interlaced with multiple disciplines. Cadres responsible for production techniques themselves must study earnestly to improve their professional skills. In addition, fish farming in paddy fields is mainly carried out by single households of farmers. It is an extremely important but difficult task to train farmers to master advanced production techniques. At the same time, due to the scattered distribution and small scale of contracted fields, it is objectively difficult for farmers to adopt advanced science and technology for integrated farming in paddy fields.

III. Shackles by Old Concepts

At present, although more than 20 provinces, autonomous regions, and municipalities in China have carried out integrated farming in paddy fields, there are not many provinces and autonomous regions that have in fact developed in a large area and formed a certain yield. The development of fish farming in paddy fields in certain provinces with large paddy fields is still very slow. From a national perspective, the development of fish farming in paddy fields is still unbalanced. The main reason is that some regions do not have enough understanding of the role and significance of developing fish farming in paddy fields, and there are two wrong ideas: One is that the excavation of ditches when farming fish in paddy fields will affect grain production and should not be strongly advocated; the other is that fish farming in paddy fields is a small-time thing, laborious and difficult to achieve a considerable yield. Therefore, it is of little significance to develop fish farming in paddy fields. At the same time, old areas of fish farming in paddy fields have been bound by traditional farming techniques for a long time, and it also requires a process to accept and master new techniques.

IV. Management Issues

The development of fish farming in paddy fields requires close cooperation and support from relevant departments. Whether agriculture, fishery, water conservancy, finance, agricultural technology, and other departments can cooperate closely is related to the smooth development of integrated farming and production in paddy fields. In addition, the water is shallow and the quantity of fish is large in terms of fish farming in paddy fields. Circumstances such as stealing fish and poisoning fish occur from time to time in some areas. There are still difficulties in fishery administration, which increase the psychological pressure on farmers of fish farming in paddy fields, de-motivate the enthusiasm of farmers in production, and also restrict the development of production of fish farming in paddy fields.

V. Lack of a Sound Service System for Fish Farming in Paddy Fields

Fish fry and fingerlings are the primary material base for fish farming in paddy fields and feed and fertilizer are the guarantees of fish farming yield. Good prevention and control of fish diseases can reduce production losses. A sound transportation and sales system for live fish can not only improve the quality of fish products but also improve the value of the products. It is very important to put great efforts into carrying out technical training, popularizing advanced science and technology of fish farming in paddy fields, and establishing a solid technical training system. In the countryside, the adjustment of fry and fingerlings, the purchase of feeds, fertilizers and pesticides, and financial loans are extremely difficult. The unsound service system of fish farming in paddy fields fundamentally restricts the development of fish farming in paddy fields.

VI. Development Countermeasures

In order to achieve unprecedented development of fish farming in paddy fields, fish farming must be put in the same important position as grain production. Fish farming in paddy fields should be considered as a strategic measure for grain production development, increasing farmers' income, and improving farmers' living standards in rice cropping regions.

1. Fully recognizing the advantages of fish farming in paddy fields

The great significance of fish farming in paddy fields should be fully understood and publicized, and the development of fish farming in paddy fields should be considered as a major measure to develop the rural economy, increase farmers' income, promote grain production with fishery, stabilize grain production with fishery, and mobilize farmers' enthusiasm for grain cultivation. Actively strive for cooperation and support from water conservancy, finance, science and technology, and other departments, combine the development of fish farming in paddy fields with the construction of farmland and water conservancy infrastructure, and promote the development of the rural economy.

The practice has proved that integrated farming in paddy fields can bring about a bumper harvest of both rice and fishery, and the benefit of output value can be doubled. This is of great significance for China, a country with less land and a dense population, to adhere to the path of connotative development, exploit production potential, increase food production, and enrich farmers. It can be predicted that the huge social demands, and high comprehensive benefits constitute a powerful driving force for the rapid development of fish farming in paddy fields, and the production of fish farming in paddy fields will certainly continue to develop at a high speed. Therefore, in order to follow the development trend and guide in a favorable direction, study and formulate preferential policies to support and encourage fish farming in paddy fields and earnestly solve technical problems and capital investment in the process of fish farming in paddy fields have become a project of "common aspiration of the people" to do practical work for farmers in rice growing areas.

2. Adjusting measures to local conditions

All localities can actively seek development according to their own resources and economic and technical conditions, adjust measures to local conditions, solve practical problems without formalism and pay attention to practical results. One-size-fits-all without considering actual conditions are not allowed, and classification and guidance should be done satisfactorily. For regions with a good basis for fish farming in paddy fields and high enthusiasm for farmers, the development direction should be in-depth. While stabilizing and expanding the area, on the premise of improving benefits, it is required to develop towards the direction of famous and special aquaculture, striving to achieve the goal of high quality and high yield of both fishery and rice.

3. Perfecting the production and service system and actively promoting advanced techniques

The perfect implementation of technical training and production services is of great importance for the development of integrated farming in paddy fields. Since 1987, advanced techniques of fish farming in paddy fields have been included in the national project of the "Bumper-harvest Programme", which has achieved significant economic and social benefits. All localities can also incorporate fish farming in paddy fields in the local "Bumper-harvest Programme" in a planned way to enhance the service quality in technology, production, and marketing. The county and township levels can determine the quantity of fry and fingerlings needed according to the planning of fish farming in paddy fields, build production and breeding sites of fry and fingerlings, and form a production and supply network of fry and fingerlings by means of centralized reproduction and decentralized breeding. Meanwhile, technical guidance on the prevention and control of diseases should be strengthened. In some regions with a concentrated area of fish farming in paddy fields, the problems of production and supply of feeds and sales of adult fish should be studied and solved in time. Purchase and sales channels should be expanded to improve the economic benefits of fish farming in paddy fields and keep farmers motivated in fish farming in paddy fields.

The practice has proved that developing fish farming in paddy fields is a good thing that serves multiple purposes and benefits the country and the people. At present, the development of China's rural economy has entered a new stage mainly characterized by the adjustment of industrial structure and the improvement of economic benefits. Therefore, it is of great significance to vigorously develop rural fish farming in paddy fields to improve the comprehensive economic benefits of paddy fields, exploit the potential of land resources and mobilize farmers' enthusiasm for production.

Chapter III High-yield Culture Techniques for High-quality Rice Machine Transplanting

Section I Growth Characteristics and Culture Characteristics of Machine-transplanted Rice

I. Growth Characteristics of Machine-transplanted Rice

1. Growth characteristics in the seedling stage

(1) Full and early seedling emergence under good temperature and humidity conditions.

All machine-transplanted rice seedlings are raised in plastic trays. After sowing, the rice trays are stacked in a centralized manner, and covered with film for heat preservation and moisturizing, which provides suitable and stable temperature and humidity conditions. Therefore, machine-transplanted rice seedlings are generally characterized by fast and even seedling emergence and high emergence rate. Generally, the seedlings emerge on the second day and fully emerge on the third day after sowing, which is 2 to 3 days earlier than conventional rice seedlings.

(2) Slow growth of the first leaf in the initial stage of seedling hardening.

Due to the centralized stacking after sowing, the seedlings are transplanted to the seedling field for seedling hardening after basically full emergence (usually 48 hours after sowing). At this stage, the first true leaf emerges, and the seedlings need a process to adapt to it. Therefore, the growth slows down and does not return to normal until the second true leaf emerges.

(3) Suitable age is the most important under the high density of machine-sowing.

With the increase of seedling age, the number of leaves increases, the leaf area increases, the light exposure of the middle and lower parts of the seedlings worsens, and the nutrient storage in the soil decreases day by day. Even worse, due to the high density of machine-sown seedlings, the nutrient area occupied by individual seedlings is relatively reduced. This contradiction between light and nutrient supply and demand leads to the inhibition of the growth of seedlings. The longer the seedling age, the poorer the quality of seedlings. Therefore, the suitable seedling age for machine-transplanting of rice seedlings with soil is 16–20 days.

2. Growth characteristics in the field stage

(1) Severe mechanical damage and long seedling recovery stage.

The foliar age of machine-transplanted rice seedlings is generally about 3.5 leaves. At this time, the seedlings are in the weaning stage, and the sowing density in the plastic tray is high. The root system between the seedlings is tightly coiled, and the root system is seriously injured during machine transplantation. The individual quality of seedlings is not as good as that of dry seedlings raised in fertilizer beds. Therefore, compared with manual-transplantation, machine-transplanted rice seedlings survive slowly and take a long time to green up.

(2) Shortened growth period and delayed growth process.

At present, machine-transplanting is mostly applied in japonica rice culture. Taking late-maturing medium-japonica rice or medium-maturing medium-japonica rice as an example, due to the limitation of seedling age and sowing period, the sowing date is generally delayed by 15–20 days compared with the conventional cultivation of the same variety, resulting in a whole growth period being 10–15 days shorter than that of conventional cultivation. With the shortening of the growth period, the heading stage and maturity stage are correspondingly delayed, so it is not suitable to select varieties with a long growth period for machine-transplanted rice.

(3) Concentrated tillering of individual plants, more seedlings at the peak of the population and reduced spike rate.

The seedling density of machine-transplanted rice is high, and the nutrient area and space of a single seedling are very small, so tillering generally does not occur in the seedling stage, and the tillering positions I and II are vacant. The incidence of tillering in positions III and IV is low in the field stage and increases with the rise of tiller position. The incidence of tillering in positions V to VII is generally 60%–90%. These tillering positions have a high incidence of tillering and a high spike rate and are the main tillering positions for high-yield culture. From the perspective of population development, the number of seedlings per unit area of machine-transplanted rice is generally more than that of conventional cultivation, but the seedlings are thin and have low dry weight. A process of thickening and weight gaining of 1.5–2 foliar ages is required for the seedlings before tillering. After tillering occurs, the population tillering accelerates rapidly. Compared with conventional cultivation, it is often about one foliar age ahead of enough seedling stage and peak seedling stage. Moreover, there are often too many seedlings in the peak seedling stage, resulting in a decrease in the spike rate and a small spike type. The spike rate of machine-transplanted rice seedlings can only reach 65%–70% in general and cannot reach the index of about 80% of manual-transplanted rice seedlings.

(4) Less differentiation of single-spike glumous flower and a high percentage of filled grains.

Because of the relatively large number of peak seedlings and high spike rate of machine-transplanted rice, the field environment in which the plants are located is relatively poor. In addition, the sowing period of machine-transplanted rice is about 15 days later than that of manual-transplanted rice, the vegetative growth period of stems and tillers of seedlings is shortened by more than 10 days, the individual development is relatively poor, and the differentiation of pedicels and glumous flowers is reduced. However, the proportion of dominant glumous flowers is slightly increased, so the percentage of filled grains is higher than that of manual-transplanted rice, which is normally 1–2 percentage points higher. However, the number of seeds per spike is still less than that of manual-transplanted rice.

(5) Machine-transplanted rice has relatively developed root systems, which are shallowly distributed but in a large rooting amount.

Since the root system is shallow in the soil, the water irrigation in the early and middle stages is shallow, and the surface air and heat of the soil are in good condition, which is conducive to rooting. With the growth process, the roots growing horizontally increase and develop to the shallow layer. In the tillering stage and spike differentiation stage, the root system of 0–5 cm layer is 12.4% and 21.7% more than that of manual-transplanted rice, respectively. In the root system distribution, the root amount in the soil layer of 0–20 cm of machine-transplanted rice accounted for more than 90%, more than 5.5% higher than that of 84.5% of manual-transplanted rice, and the proportion of manual-transplanted rice below 20 cm was significantly more than that of machine-transplanted rice. The root system of machine-transplanted rice is relatively developed, widely and deeply distributed, which enhances the absorption and transportation function of water and nutrients of rice plants, thus promoting the solid stem, thickening of leaves, prolonging the life of roots and leaves, and facilitating the increase of yield.

II. Culture Characteristics of Machine-transplanted Rice

1. Short seedling stage and long field stage

The carpet seedlings used for machine-transplanted rice are generally 15–18 days old, and their foliar age is generally 3–4 leaves younger than that of conventional wet seedlings. Therefore, the effective tillering position of machine-transplanted rice in the large field is increased by 3–4. The characteristics of a longer effective tillering stage should be used to strive for early tillering and improve the tillering spike rate.

2. Thin seedlings with poor stress resistance

Under dense planting conditions, the floor area of a single seedling is about 0.6 cm^2. The seedlings are dense and thin, and their resistance to adverse conditions such as sun exposure, flooding, and drought is poor. Therefore, cultivating standard seedlings and enhancing stress resistance are emphasized .

3. Large machine-transplanting spacing with a certain hole missing rate

Machine-transplanted rice generally adopts the culture mode of "wide row spacing and narrow plant spacing". For example, the row spacing of TYM series rice transplanters is fixed at 30 cm. Such row spacing is too large for some multi-spike japonica rice varieties, so the plant spacing should be controlled to ensure appropriate density. Machine-transplanted rice inevitably has a certain amount of missing transplantation. If the hole missing rate is too high, it may cause a large number of missing plants and affect the yield. Effective measures should be taken to control the hole missing rate within the allowable range.

4. Late sowing and late growth period

Machine-transplanted rice has a delayed transplanting period, young seedlings, and a lagging growth process compared with conventional cultivated rice, and cannot be equivalent to conventional cultivated rice in fertilizer operation. The specific dates of disease and pest control and water management, including the time of field draining and water cut-off, are correspondingly later than those of conventional rice cultivation.

Section II Seedling Raising Technique for Machine-transplanted Rice

I. Carpet Seedling-raising Technique

At present, the carpet seedling-raising technique is widely used in machine-transplanted rice in Jiangsu. Soft trays are basically used for carpet seedling raising. This technique is a low-cost simplified seedling-raising method summarized from the practice of industrialized seedling raising. It is simple, convenient, low-cost, of good quality, and has a high success rate.

1. Operation process of carpet seedling raising (Fig. 3-1)

2. Preparation of bed soil

(1) Selection of bed soil.

Fertile loam with no residues, gravels, weeds, and pollution should be selected. The suitable bed soils include: first, vegetable garden soil; second, cultivated and matured dry field soil (it is not suitable to take soil from wild grassland and wheat fields sprayed with herbicide in the current season); third, paddy field soil after autumn plowing, winter harrowing, and spring plowing. At present, substrate

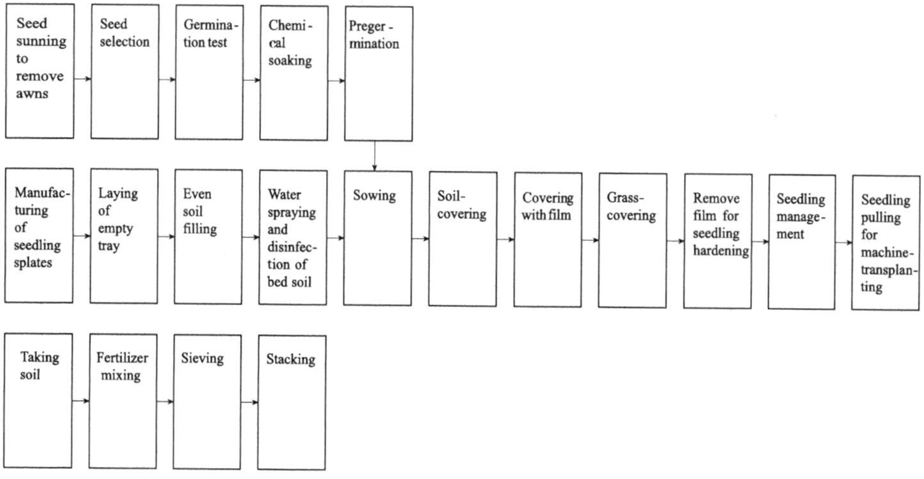

Fig. 3-1 Operation Process of Carpet Seedling Raising

seedling raising is adopted in many places. The seedling substrate is used to replace nutrient soil (subsoil), which is convenient and simple, with strong seedling moisturizing ability, good packing, light seedling blocks, and easy rolling and transportation.

(2) Bed soil consumption.

Generally, 100 kg of fine nutrient soil is required for each mu of field as bed soil, and 25 kg of unfertilized fine-sieved soil is required as seed-covering soil.

(3) Bed soil fertilization.

Fertile and loose garden soil can be directly used as bed soil after sieving. For other suitable soils, it is recommended to take soil in winter. Before taking soil, fertilization is generally required for the land. 2,000 kg of decomposed human and animal manure (no plant ash), 60–70 kg of 25% nitrogen, phosphorus, and potassium compound fertilizer; or 30 kg of ammonium sulfate, 40 kg of calcium superphosphate, 5 kg of potassium chloride, and other inorganic fertilizers are evenly applied per mu. It is advocated that seedling hardening agent suitable for local soil properties be used to replace inorganic fertilizer, and 0.5–0.8 kg of seedling hardening agent be mixed for every 100 kg of fine soil during bed soil processing and sieving. Calcium superphosphate can be added as appropriate to reduce the pH value (the appropriate pH value is 5.5–7.0) when the pH value of the land is relatively high. After application, machine rotary tillage is carried out continuously 2–3 times, and the topsoil is piled up and covered with agricultural film until the bed soil is matured.

(4) Bed soil processing.

Screening is carried out in sunny weather and when the moisture content of the soil heap is suitable (the moisture content is 10%–15%, the fine soil is kneaded into balls by hand, which is scattered when it falls to the ground). The particle size of fine soil should not be greater than 5 mm, and the soil particles with a particle size of 2–4 mm should reach more than 60%. After the sieving, continue to stack the soil and cover it with agricultural film in a centralized manner to promote the full maturation of fertilizer and soil. The crusher can also be used for soil crushing.

For early rice seedling raising and in areas prone to coldness in late spring, in order to prevent seedling diseases such as seedling blight, 50–60 g of 65% Dexon mixed with 1,000–1,500 times water should be applied per cubic meter of bed soil for disinfection.

(5) Advocating, paying attention, and prohibiting.

It is advocated that fertilizer be increased before winter, and that pH adjustment and disinfection should be carried out for alkaline soil. It is advocated to use seedling hardening agent, pH adjustment, disinfection, chemical control and fertilization. Add 0.5–1.0 kg of seedling hardening agent per 100 kg of fine soil and mix them well.

Attention should be paid to the prevention of seedling stage diseases such as seedling blight, and it is particularly necessary to apply Dexon for bed soil disinfection, especially in areas prone to low temperatures in spring and coldness in late spring. Attention should be paid that soils that have not been fertilized before winter should not be fertilized, because top dressing during the weaning stage can also cultivate strong seedlings. If fertilization is really necessary, it should be carried out at least 30 days before sowing. After being mixed with fertilizer and sieved, the soil must be covered with a film to promote decomposition.

It is prohibited to directly mix undecomposed stable manure, sludge, urea, ammonium bicarbonate, and other fertilizers to prevent fertilizers from burning seedlings. It is prohibited to use fertilized nutrient soil as seed-covering soil.

3. Seed preparation

(1) Variety selection.

Superior varieties with high quality, high and stable yields, medium tillering, good resistance, and a balanced number of spikes/grains that are suitable for local planting should be selected according to different cropping, variety characteristics, and safe full heading stage. Under the same conditions, it is advisable to select the varieties with a relatively short growth period.

(2) Seed consumption and sowing density in fields.

The consumption of hybrid rice seeds is 1–1.5 kg/mu and 3–3.5 kg/mu for conventional japonica rice.

In order to meet the requirements of machine transplanting, the sowing amount of machine-transplanted seedlings is relatively high. Generally, the sowing amount of each tray of bud grain of hybrid rice is 80–100 g, and the sowing amount of bud grain of conventional japonica rice is 120–150 g. Too-high or too-low sowing amount is not conducive to cultivating qualified rice seedlings for machine transplanting. The most basic principle for determining the appropriate sowing density is uniformity and packing, that is, referring to the number of seedlings planted in each hole in the large field. On the premise of ensuring uniform sowing and the root system of seedlings can be coiled, the sowing amount can be appropriately reduced according to the variety, climate, and other factors so as to improve the quality of seedlings and increase the flexibility of seedling age. The calculation formula for a suitable sowing amount is:

Sowing amount (dry seeds) = [Actual number of seedlings/(Germination rate/ Seedling rate/1,000)] × Thousand-seed weight

Taking hybrid rice with a germination rate of 97% and a thousand-seed weight of 26 g as an example, the seedling rate under high-density seedling-raising conditions was roughly 85%. Generally, it is required that the number of final seedlings of hybrid rice should be 1–1.5 per cm^2. If calculated as 1.5, there should be 2,436 seedlings per plate of 1,624 cm^2, equivalent to the sowing amount of about 100 g bud grains. In general, the sowing amount of early rice can be increased appropriately in the range of 80–100 g, while that of middle and late rice can be decreased appropriately.

According to the experience of recent experiments and large-scale production in Jiangsu Province, the suitable sowing density for medium- and late-maturing japonica seedlings in plastic trays with two crops in a year is 230–250 seeds/dm^2. For example, the thousand-seed weight of seed grains is 27 g and the germination rate is 95%, equivalent to about 920 g of bud grains per m^2 and 150 g per tray. The suitable foliar age of seedlings with this sowing amount is 3.1–3.9 leaves, with a certain flexibility of seedling age, which can reduce the risk of exceeding the seedling age and seriously affect large field production.

(3) Seed treatment.

The seed consumption, seed soaking, and pregermination time should be calculated in advance according to the sowing date and machine-transplanting area.

① Determination of the sowing date. There are significant differences between machine-transplanting seedling raising and conventional seedling raising: First, the sowing density is high; second, the root system of seedlings is concentrated in a thin soil layer with a thickness of only 2–2.5 cm, so the flexibility of seedling age is low. The sowing date must be calculated according to the cropping pattern and the seedling age is about 20 days. The fields can wait for seedlings, but the seedlings can't wait for the fields. If the machine-transplanting area is large, the transplanting progress should be arranged according to the working efficiency of the rice transplanter and the proficiency of the machine operator, and the seeds should be soaked in batches and sown in sequence to ensure that the seedlings are transplanted at the appropriate age. At present, rice/(barley and wheat) and rice/rape cropping are the main cropping in Jiangsu Province. Under different cropping conditions, the suitable sowing date is roughly shown in Table 3-1.

Table 3-1 Suitable Sowing Date for Machine-transplanted Rice with Different Cropping in Jiangsu Province

Region	Cropping	Transplanting stage (early planting)	Transplanting in 3-leaf stage	Transplanting in 4-leaf stage
South Jiangsu	Rape (barley)	May 25–30	May 10–15	May 5–10
	Wheat	June 5–10	May 20–25	May 15–20
Central Jiangsu	Rape (barley)	May 25–30	May 10–15	May 5–10
	Wheat	June 10–15	May 25–30	May 20–25
North Jiangsu	Rape (barley)	June 1–5	May 15–20	May 10–15
	Wheat	June15–20	May 30–June 5	May 25–30

② Selection of seeds. Standard commercial seeds should be selected as far as possible. Ordinary seeds should be properly subject to sunning, awn removal, selection, and germination tests before soaking. The germination rate of seeds is required to be above 90% and the germination potential to be above 85%. When the traditional saline method is used for seed selection, the specific gravity of water is 1.06–1.10 (that is, when fresh eggs are put into the saline, the exposed area is equal to the size of a 1-yuan coin). After the seed selection with saline, the seeds should be washed with clean water to remove the salt outside the grain and prevent it from affecting germination. After washing, the seeds should be dried for later use or directly soaked. Hybrid rice seeds are generally selected by air separation. Before seed selection, the seeds should be sunned for 1–2 days, and then the hollowed seeds should be blown away with low air volume to ensure that the seeds are uniform and full and have strong germination potential.

③ Chemical soaking. The diseases of rice mainly carried by rice seeds include rice bakanae, rice blast, rice false smut, leaf blight, as well as rice stripe disease transmitted by small brown planthopper in the seedling stage, which can be controlled by chemical soaking. When soaking the seeds, add 10 g of imidacloprid to 1 tube (2 mL) of Shibaike (prochloraz) and 6–7 kg of water to soak 5 kg of seeds. The soaking time depends on the temperature. Generally, japonica rice needs to be soaked for about 3 days and indica rice for about 2 days. The criteria when rice seeds absorb enough water is that the husk is transparent, the belly of the rice grains is white, and the rice grains are easy to break without sound.

④ Pregermination. The main technical requirements for pregermination are "fast, neat, uniform, and strong": "Fast" means to complete pregermination within 2 days; "Neat" means the requirement of germination potential higher than 85%; "Uniform" means the neat and uniform bud length; "Strong" means the buds are strong, with an appropriate ratio of root to bud length, bright white color, fragrance, and no alcoholic odor. According to the main process and characteristics of seed growth and germination, pregermination can be divided into three stages: high-temperature chest breaking, appropriate-temperature accelerating pregermination, and air-drying hardening.

A. High-temperature chest-breaking. The period from the stacking of the rice to the time when the seed embryo breaks through the grain shell is called the chest-breaking stage. Enough water and a suitable temperature are the main requirements for fast and neat chest-breaking. Within the upper-temperature limit of 38°C, the higher the temperature, the more vigorous the physiological activities of the seeds, and the faster and neater the chest-breaking, and vice versa. Generally, after the temperature of the rice stack rises, the temperature of the upper, lower, inner, and outer sides of the stack should be consistent. If necessary, plowing and mixing should be carried out to ensure the uniform temperature of the rice seeds and promote neat and rapid chest-breaking.

B. Appropriate-temperature pregermination. The period from the chest breaking of rice seeds to the time when the elongation of young buds meets the sowing requirements is called pregermination stage. The criteria for pregermination of double-membrane manual-sown seedlings is that the root length is up to 1/3 of the seed and the bud length is 1/5 to 1/4. If machine sowing is adopted, it is okay when 90% of the seeds break their chest and become white. Buds grow in wet conditions, while roots grow in dry conditions. The length of roots and buds is mainly controlled by adjusting the moisture content of rice. At the same time, the temperature of the grain pile should be adjusted in time to maintain the temperature at the pregermination stage at 25–30°C, so as to ensure the coordinated growth of roots and buds and the thickness of roots and buds.

C. Air-drying hardening. In order to enhance the adaptability of bud grain to the external environment after sowing and improve the uniformity of sowing, air-drying hardening is also required after pregermination. Generally, after pregermination, the seeds are spread indoors for drying 4–6 hours, and when the seed moisture is suitable and not sticky to hands, they can be sown.

4. Model and specification of the seedling tray

When raising seedlings with soft trays, about 25 soft trays are generally prepared for each mu of field. If the machine sowing line is adopted, each line needs to be equipped with sufficient hard trays for turnover. The surface of the soft trays should be flat and smooth, with vertical edges and corners, and without color difference, wrinkles, distortion, incompleteness, cracks, and burrs.

5. Seedbed treatment

According to the scale of seedling raising, choose the mature land with flat terrain, separated drainage, and irrigation, and clean water sources which are leeward, sunward, and adjacent to fields as the seedling field, or choose the fertile vegetable field near the house. The ratio of seedling fields to large fields should be 1 : 80–1 : 100, and generally, 7–10 m^2 seedling field is required per mu. Wild grassland, spray-injury farmland, and soil-water-polluted farmland should not be selected.

Seedling plates are prepared 10 days before sowing. The seedbed is 1.4–1.5 m wide, and the length is determined according to the needs and the size of the plot. A drainage ditch and management channel with a width of 20–30 cm and a depth of 20 cm is reserved between the seedling plates. The peripheral ditch around the seedling field is 50 cm deep, and the ridge is flat and compact. The ridge surface is generally 15–20 cm higher than the seedling bed, with a leveling gap reserved (Fig. 3-2). In order to make the surface of the seedling plate flat, water can be introduced first for leveling, and then the water is drained to dry the seedling plates and make the plate surface solid. Two days before sowing, shovel the high place and fill the low place, fill the cracks, and fully compact them to make the plate surface compact, flat, smooth, and straight (Compact: the seedling plates should be solid without sinking foot; Flat: the plate surface should be flat; Smooth: the plate surfaces should be free of residues and sundries; Straight: the template is neat and the ditch edge is vertical).

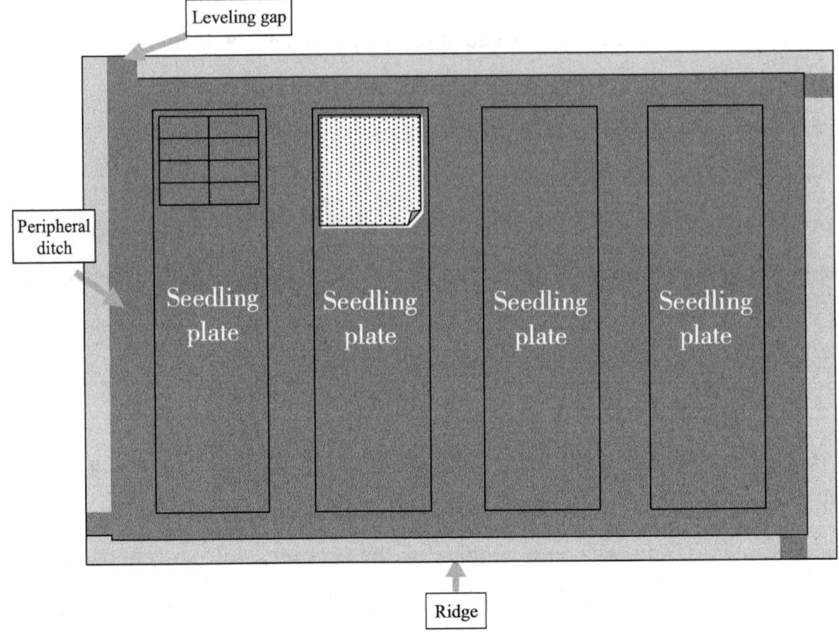

Fig. 3-2 Seedbed Treatment

6. Precision sowing

(1) Manual sowing.

In the process of soft tray seedling raising, the standardization of operation links is the basic requirement to ensure the quality of seedling raising. Among them, the sowing quality is directly related to the quality of seedlings and the quality of machine transplanting. Therefore, in actual operation, the sowing amount should be accurately calculated according to the specific variety, and efforts should be made to sow evenly.

① Sequential laying of trays. Soft trays are laid horizontally on the seedling plates. In order to make full use of the seedling plates and facilitate the seedling pulling, each seedling plate has two rows of trays laid horizontally in a tight and neat manner. The flashes between the trays should be overlapped, and the bottom of the trays should be closely attached to the bed surface.

② Even spread of bed soil. Spread the prepared bed soil with a uniform thickness of 2–2.5 cm, and flatten the soil surfaces.

③ Water replenishment and soil moisture conservation. One day before sowing, irrigate the ditch water and drain it quickly after the bed soil is fully moisturized, or directly sprinkle water with a watering can before sowing. When sowing, it is required that the soil-saturated moisture content reach 85%–90%. Together with watering the bottom water before sowing, 1,000–1,500-fold dilution of 65% Dexon can be sprayed on bed soil for disinfection.

④ Sowing seeds are weighed by tray when sowing. Generally, 120–150 g of conventional japonica rice and 80–100 g of hybrid rice in each tray are sown evenly after chest breaking and whitening. In order to ensure uniform sowing, 4–6 trays can be used as a group for sowing, and fine sowing several times is required to ensure uniform sowing.

⑤ Spread the covering soil evenly. After sowing, spread the seed-covering soil evenly, with a thickness of 0.3–0.5 cm. The covering soil should not be too thick and should just cover the bud grains. Pay attention to the use of sieved fine soil without fertilization, and do not use nutrient soil mixed with seedling hardening agent. Do not sprinkle water after seed-covering soil is spread to prevent the hardening of topsoil from affecting the emergence of seedlings.

⑥ Film covering and soil moisture conservation. After covering the soil, fill the ditch with water to wet the seedling plates and then discharge it quickly to make up for the insufficient moisture of the seedling plates, and trim the plate edges along the periphery of the seedling plates to ensure the size of the seedling blocks. After sowing, bud grains need to undergo a certain high temperature and humidity to achieve neat seedling emergence. Generally, the temperature is required to be 28–35°C and the humidity to be above 90%. Therefore, after sowing and soil covering, the seedlings should be covered with film and grass to control temperature and maintain moisture for full emergence. Before film covering, a fine reed or a thin layer of wheat straw should be placed on the plate surface every 50–60 cm to prevent the bed soil from being stuck to the agricultural film and causing seed stuffiness. The surroundings of the agricultural film must be sealed tightly and covered with a layer of straw. The thickness of the straw should be such that the agricultural film can be seen, so as to prevent the high temperature from burning the buds at noon on sunny days. For early spring seedling raising with low temperatures or in areas prone to coldness in late spring, a tunnel should be built on the basis of film covering to increase temperature and raise seedlings. The tunnel is about 0.45 m high, the spacing between arch frames is 0.5 m, and the surroundings should be tightly sealed after film covering. In the area where rodent damage occurs, rodenticides should be sprayed around the seedbed film, and do not spray rodenticides into the tunnel film.

(2) Mechanized seedling raising and sowing.

Mechanical sowing can also be used for tray seedling raising. The rice seedling machine can carry out soil paving, seed sowing, soil covering, and other operations, respectively, while the sowing line can complete the above processes at one time. Machine sowing has high efficiency and good quality. At present, many seedling sowing machines are used in production, mainly including hand-push seeders, hand-crank seeders, rice seedling raising and sowing lines, etc. Their basic principles are the same. By seeding methods, they are mainly divided into two types: strip seeding and indented seeding. The sowing machine with a strip seeding mechanism has high operation efficiency, but the uniformity of sowing is not ideal under the condition of a low sowing amount. While the problem of uniformity under the condition of a low sowing amount is basically solved by the indented sowing method, its operation efficiency is relatively low.

When machine sowing is adopted, an appropriate number of hard plastic trays should be prepared for turnover. Before sowing, the seeder should be commissioned to stabilize the thickness of the subsoil in the tray at 2–2.5 cm. The sowing amount should be accurately adjusted to stabilize each tray of sowing buds within the predetermined suitable range. The thickness of covering soil is 0.3–0.5 cm, and it is appropriate to cover bud grains completely. The watering amount should be controlled until the subsoil moisture reaches the saturated state. The dry soil on the plate surface should naturally absorb moisture without a white surface within 10 min after covering with the soil.

After sowing, the hard trays can be removed in the field and soft trays can be placed on the seedling plates. Alternatively, the seedling plate can be stacked indoors for heating and budding, and then moved to the seedling field to remove the trays. At this time, the tight and neat soft trays still need to be laid horizontally in two rows on the seedling plates in turn, and the bottom of the trays are closely attached to the bed surface. After sowing, cover the seedlings with non-woven fabrics immediately and keep the non-woven fabrics wet during the seedling stage. The non-woven fabrics loosen freely with the growth of seedlings, and should be removed in time according to the growth of seedlings.

7. Seedling stage management

Cultivating strong seedlings suitable for machine transplanting is the key to the success or failure of promoting mechanized rice transplanting. As the saying goes, "Good seedlings promise a half-harvest and high yield"; the quality of rice seedlings plays an important role in the number of spikes, number of grains, and grain weight in the later stage of rice growth. The basic requirements of mechanized rice transplanting for rice seedlings are overall balance and strong individuals, and it is required that the height of rice seedlings in one plate is the same, and the thickness of rice seedlings in one handful of rice shoots is the same. Therefore, seedling stage management is more technical and normative.

(1) High temperature and humidity for full emergence.

The pregerminated rice seeds need to go through a high-temperature and humidity seedling establishment stage after sowing to ensure neat seedling emergence. Therefore, corresponding heating and moisturizing measures should be taken according to different seedling-raising methods and cropping to ensure safe and full emergence. At the same time, the seedling field should have a leveling gap to prevent the rain from flooding the seedbed and causing seed stuffiness and rotten buds.

Film and grass covering for seedling establishment is suitable for rice seedling raising when the temperature is high. It includes three types: double-film seedling raising, soft tray manual sowing, and machine sowing direct tray removal.

Attention should be paid to two points in the seedling establishment stage: First, the thickness of the grass cover should be controlled and uniform so as to avoid burning seedlings at noon on sunny days; second, after rain, the accumulated water on the covering film should be removed in time to avoid water accumulation on the film. In addition, the wetted covering straw will become heavier and the local part is pressed, resulting in seed stuffiness and rotten buds and affecting the whole seedling.

Tunnel seedling establishment. The method is suitable for early spring seedling raising with low temperatures or areas prone to coldness in late spring. After the top buds emerge from the soil surface, the surface temperature in the tunnel at noon on sunny days should be controlled below 35°C to prevent high temperatures from burning the seedlings.

The period from sowing to seedling emergence is generally the film-sealing stage in the tunnel, which mainly needs heat preservation and moisturizing. Only when the temperature in the film exceeds 35°C can both ends of the seedbed be uncovered at noon for ventilation and cooling, and then covered in time. If the bed soil turns white and the seedlings roll leaves during this period, alternant water should be irrigated to maintain moisture.

(2) Timely seedling hardening.

Uncovering film for seedling hardening. The seedlings should not be covered for too long, and the uncovering of the film depends on the temperature at that time. Generally, the film uncovering for seedling hardening is carried out when the seedlings emerge from the soil for about 2 cm and the first leaf is incomplete until the first leaf comes out (3–5 days after sowing). If the covering time is too long, the hot sun makes it easy to burn the seedlings. Principle of film uncovering: The film can be uncovered in the evening on sunny days, in the morning on cloudy days, before light rain, and after heavy rain. In the case of cold currents and low temperatures, it is advisable to delay the film uncovering and to uncover the film during the day and cover the film at night.

Seedling hardening in tunnel. After the seedlings turn green, the time for removing the tunnel should be determined depending on the temperature. When the minimum temperature is stable above 15°C, the tunnel can be removed. Otherwise, the method of uncovering the film during the day and covering the film at night has to be adopted for management, so the soil (or bed soil) can be kept moist.

(3) Scientific water management.

① Moisture management. Intermittent irrigation is adopted to create a mainly humid environment, thus achieving the purpose of regulating air, fertilizer, and temperature and protecting seedlings with water. Key points of operation: When removing the film, fill the ditch with water, and fill water again after natural drying. Repeat the above steps. Irrigate seedlings with thin water to protect them if leaf rolling occurs at noon on sunny days, and drain the water in the seedling ditch on rainy days. In the case of strong cold air invasion in early spring, irrigate the seedlings with water to half the height of the plants to protect them. After warming up, change the water to protect them after the temperature is stable, so as to prevent root damage caused by low temperature as well as rotten seedlings and dead seedlings caused by excessive temperature difference. After the temperature is normal, drain and ventilate in time to improve the root system vitality of seedlings. Water control should be carried out for seedling hardening 3–5 days before transplanting.

② Water control management. It is basically similar to the water management technique for dry seedling raising in conventional fertilizer beds; that is, sufficient water is irrigated once when the film is removed (the water level is flush with the ditch), and then drained after the bed soil is fully soaked (water supplement by spraying is also optional). At the same time, clean the seedling ditches, keep the water system unblocked, ensure that there is no accumulated water in the seedling fields on rainy days, and prevent dry seedlings from being flooded and losing the advantage of dry raising. After that, if the leaves of the seedlings roll at noon, spray water manually once in the evening or the early morning of the next day to moisten the soil. If there is no leaf rolling, no water supplement is needed. The quality of water for supplements should be clean; otherwise, it is easy to kill seedlings.

(4) Make good use of weaning fertilizer.

Weaning fertilizer should be applied according to the specific conditions such as bed soil fertility, seedling age and temperature, generally in the one-leaf-one-bud stage (7–8 days after sowing). Mix 500 kg of decomposed manure with 1,000 kg of water or 5 kg of urea (about 2 g of urea per tray) with 500 kg of water per mu of the seedling pond field, and apply them on the seedling leaves in the evening at the time of guttation. They may also not be applied for fertile bed soil. In order to prevent the seedlings from growing too high in the wheat cropping field, the amount of fertilization can be appropriately reduced.

(5) Disease and pest control.

The main diseases and pests in the seedling field stage include rice thrips, small brown planthopper, seedling blight, borer, etc. Close attention should be paid to the occurrence of diseases and pests in the seedling field stage, and timely symptomatic pesticides should be used for control. In recent years, the occurrence of rice stripe disease has aggravated year by year, so it is necessary to prevent and control small brown planthopper, which can be sprayed with imidacloprid 2g (active ingredient) plus 80 kg water at the one-leaf-one-bud stage. In addition, the temperature is low and the temperature difference is large during the seedling raising in early spring, which makes the seedlings vulnerable to seedling blight. After the film is removed, the seedling bed is supplemented with water, and 600–750 kg of 1,000–1,500-fold dilution of Dexon is sprayed on each mu of the seedling pond for prevention.

(6) Auxiliary measures.

While improving the sowing quality and managing fertilizer and water in the early and middle stages of the seedling field, strengthening agents can be applied in accordance with the weather and seedling growth during the second leaf stage. If the temperature is high and the rain is too much, the seedlings grow rapidly, especially for seedlings that cannot be transplanted at the right time, spray 50 g of 2,000-fold dilution of 15% paclobutrazol wettable powder per mu of the seedling pond field (do not apply too much or spray ununiformly. If the "dry seedling strengthening agent" has been applied during bed soil fertilization, it is not necessary to apply it) to delay the growth of plants. At the same time, it promotes lateral growth and increases the dry matter content of seedlings.

8. Preparation before transplanting

Preparation before machine transplanting of seedlings is an important measure to ensure the quality of seedlings, enhance stress resistance after planting and promote the early growth and rapid development of seedlings. The following aspects should be focused on.

(1) Apply transplanting fertilizer according to seedling condition.

If the seedlings have high nitrogen levels, they also have strong rooting ability, high carbon levels, and strong transplanting injury resistance. In order to make the seedlings have strong rooting ability and strong resistance to transplanting injury during transplanting, it is necessary to observe the seedlings' condition and apply transplanting fertilizer before transplanting, so as to promote the seedlings to be dark green and the leaves to be vigorous and delicate. The specific fertilization should be carried out in batches according to the progress of machine transplanting, generally 3–4 days before transplanting. The amount of fertilizer and application method should depend on the color of the seedlings. For seedlings with faded-color leaves, 4–4.5 kg of urea mixed with 500 kg of water per mu should be evenly sprayed or poured in the evening. After application, clear water should be sprayed to prevent fertilizer from burning the seedlings. If the leaves are normal in color and straight without dropping, spray 1–1.5 kg of urea mixed with 100–150 kg of water on the roots per mu. If the leaves are dark green and dropping, do not apply fertilizer, and take water control measures to improve the seedling quality.

(2) Timely water control and seedling hardening.

Before transplanting, water control and seedling hardening are important means to reduce the free water content in the seedlings, improve the carbon level and enhance the stress resistance of seedlings. The water control time should be determined according to the weather conditions before transplanting. Because of early sowing and early transplanting of spring seedlings, the temperature, lighting intensity, and transpiration of seedlings are relatively low before planting when compared with wheat cropping seedlings. Generally, water control and seedling hardening are carried out 5 days before transplanting. The temperature is high and the transpiration is large before transplanting wheat cropping seedlings, so the water control should be carried out 3 days before transplanting. Water control method: Keep half of the ditch water on sunny days. If the seedlings roll leaves at noon, sprinkle the water to supplement the water. On rainy days, the accumulated water in the seedling ditches should be drained, especially before seedling pulling and transplanting. Before the rain, the seedlings should be covered with film to prevent the high moisture content of the bed soil from affecting seedling pulling and transplanting.

(3) Apply pesticides before transplanting.

Because the machine-transplanted seedlings are small and tender, they are vulnerable to borer, rice thrips, and rice weevil after transplanting. Chemical control should be carried out before transplanting. Spray 30–35 mL of 2.5% Phoxim emulsifiable concentrate mixed with 40–60 kg of water per mu 1–2 days before transplanting. In the area where rice stripe disease occurs, 15 mL of 10% imidacloprid emulsifiable concentrate should be applied per mu to control the virus-carrying transmission hazard of small brown planthopper, so as to achieve transplanting with pesticide and treat several diseases with one pesticide.

II. Potted Seedling-raising Technique

Potted seedling machine transplanting is a major technical innovation in rice production mode, and a new technique integrating the advantages of rice seedling throwing and machine-transplantation. Generally, its yield is increased by about 10% when compared with carpet seedling machine transplanting in the plastic tray. The bowel tray can raise seedlings with complete pot-shaped nutrient soil blocks at the root, which are characterized by low-density sowing, long seedling age, large dry mass, and high fullness. Potted seedlings are transplanted with pot soil without damage to roots and transplanting injury. Therefore, there is basically no seedling recovery after transplanting; the tillering occurs early, the seedlings grow vigorously in the early stage, the plants are thick and strong, the spikes emerge early and are large with many grains, and the yields are high. Machine transplanting of potted seedlings is beneficial to the early ripening of rice, which is of certain significance to ensure the timely sowing of the next crop of wheat in multi-cropping areas. At present, the potted seedling-raising technique is adopted in some large farms. It is also a technique actively promoted by the state.

1. Operation process

Seedling preparation (variety selection and preparation, bed soil fertilization and processing, seedling field production, material preparation) → precision sowing → dark treatment for seedling emergence → tray placement → seedling field management → large field machine transplanting → large field management.

2. Preparation of bed soil

(1) Selection of bed soil.

The method is the same as that of carpet seedling raising.

(2) Bed soil consumption.

Sufficient nutrient soil is prepared for conventional rice at 70 kg/mu, and sufficient nutrient soil is prepared for hybrid rice at 60 kg/mu, with a soil consumption of about 1.5 kg per tray.

(3) Bed soil processing.

Compared with carpet seedlings, the mechanized sowing of rice by potted seedling machine transplanting requires two passes of compaction on nutrient soil, resulting in too-tight soil and poor water & air permeability. Rice seeds are prone to lack oxygen during germination, resulting in rotten buds and seeds of rice, which seriously affects seedling emergence. Therefore, solving the contradiction between soil pelletization and soil aeration is the key to improving the seedling rate of potted rice seedlings. By adding seedling-strengthening agent, river sand, and soil binder during the preparation of nutrient soil, the porosity of the soil can be effectively adjusted to keep the soil in the pots with suitable water content, no hardening, and good ventilation, thus creating favorable conditions for rice germination and seedling emergence. The seedling-strengthening agent plays the role of fertilizer supply, acid adjustment, height control, disease prevention, etc. Add 0.5 kg seedling-strengthening agent to every 100 kg of fine soil, strictly control the amount, and fully mix with nutrient soil to prevent seedlings and buds from being injured by the ununiform application of the seedling-strengthening agent. The river sand can obviously improve the soil permeability and the compressive strength of the potted soil ball. The amount of river sand should be more than 30% of the total weight of nutrient soil. Soil binder has the functions of promoting root system growth, improving seedling packing, improving soil agglomeration, increasing the porosity of soil crumb, and improving water permeability. Soil binder is used at 50g per mu. The above three seedling-raising materials of the seedling-strengthening agent, soil binder, and river sand are evenly mixed with the sieved fine soil 3 days before sowing and then piled up for later use.

3. Seed preparation

(1) Variety selection.

Superior varieties with high quality, high and stable yields, medium tillering, good resistance, and a balanced number of spikes/grains that are suitable for local planting should be selected according to different cropping, variety characteristics and safe full heading stage. Under the same conditions, it is advisable to select the varieties with a relatively short growth period.

(2) Seed consumption and sowing density in fields.

The potted seedlings for machine transplanting have a small sowing amount, large planting row spacing, and a small number of basic seedlings. Generally, the number of basic seedlings of hybrid rice is 25,000–30,000/mu, and the number of basic seedlings of conventional rice is 60,000–70,000/mu, which are reduced by 15,000–20,000/mu and 30,000–40,000/mu respectively compared with the carpet seedlings for machine transplanting. The way to increase the yield of rice transplanted by the potted seedling machine is based on the appropriate number of spikes and relying on strengthening stalks and enlarging spikes. In general, the amount of seeds used in large fields is about 3.0 kg/mu for conventional japonica rice, about 2.0 kg/mu for hybrid japonica rice, and about 1.5 kg/mu for hybrid indica rice. Before sowing, the seed awns must be mechanically removed, and the seed quality should meet the standard *Seed of food crops—Part 1: Cereals* (GB 4404. 1–2008). The number of single-hole seedlings is very important for tillering and yield.

The seed amount of conventional japonica rice is about 3.2 kg/mu, with an average of about 4 seedlings per hole, 6–7 seeds per hole, and about 80 g of dry seeds per tray. The seed amount of hybrid japonica rice is about 2.25 kg/mu, about 3 seedlings per hole, 4–5 seeds for each hole, and the suitable sowing amount per tray is about 50g. The seed amount of hybrid indica rice is about 1.5 kg/mu, about 2.5 seedlings per hole, 3–4 seeds for each hole, and the suitable sowing amount per tray is about 40 g. The number of hard trays is prepared according to 40 seedling trays/mu for conventional rice, 35 seedling trays/mu for hybrid japonica rice, and 30 seedling trays/mu for hybrid indica rice.

(3) Seed treatment.

The seed consumption, seed soaking, and pregermination time should be calculated in advance according to the sowing date and machine-transplanting area.

① Determination of sowing date: The transplanting date is determined according to the characteristics of the rice variety, the safe full heading stage and the cropping. Generally, the sowing period should be inversely calculated based on a suitable transplanting date and an assumed seedling age of 20–25 days, and the sowing should be carried out in stages to prevent over-age transplanting.

② Selection of seeds: The method is the same as that of carpet seedling raising.

③ Chemical soaking: The method is the same as that of carpet seedling raising.

④ Pregermination: The method is the same as that of carpet seedling raising.

4. Model and specification of the seedling tray

The potted seedling tray is the core carrier of the complete set of rice machinery. It is injection molded with special resin. The bottom of the seedling tray is evenly distributed with Y-shaped petal holes that can be freely opened and closed. It has both drainage and nutrient absorption functions of the seedbed, featuring strong elasticity, high flatness, aging resistance, long service life (more than 8 years), green and environmental protection, etc. See Table 3-2 for the main technical parameters. The seedling tray can raise seedlings with complete potted-shaped nutrient soil blocks at the root. The seedlings are characterized by an independent pot, developed root system, heavy seedling dry weight, and high fullness. The seedlings should be transplanted with potted soil without damage to roots or transplanting injury, and there should be no recovery after planting. They will grow vigorously, which is conducive to cultivating strong, high-quality, and environmentally friendly high-yield populations.

The number of hard trays is prepared according to 40 seedling trays/mu for conventional rice, 35 seedling trays/mu for hybrid japonica rice, and 30 seedling trays/mu for hybrid indica rice.

Table 3-2 Main Technical Parameters of Potted Seedling Tray

Name		Rice potted seedling tray
Model		D448P
Overall dimensions	Length (mm)	618
	Width (mm)	315
	Height (mm)	25
Mass (g)		420
Pot type	Upper part (mm)	Φ16
	Lower part (mm)	Φ10
	Petal shape	Y-shaped
	Number of pots (Nr.)	448

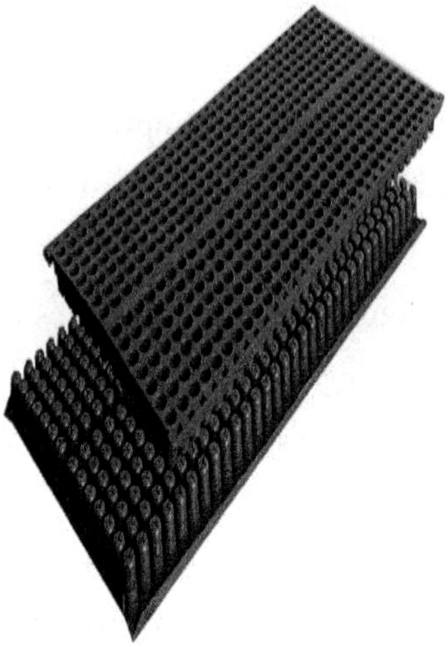

5. Seedbed treatment

Select dryland with flat plots, fertile soil, convenient seedling transportation, and good irrigation and drainage conditions. It is desirable that the ratio of seedling fields to large fields for conventional rice be 1 : 50, and that for hybrid rice be 1 : 60. The seedling fields must be plowed and sunned in advance for crushed soil. Practice has proved that soil preparation and seedling plate making for fields transplanted by a potted seedling machine should not be done by water soil preparation, but by dry soil preparation. The main reason is that the surface soil of the seedling field upon water soil preparation is prone to cracking and tilting after dehydration and drying, resulting in an uneven surface of the seedling plate. The seedling tray cannot be closely attached to the seedling plate, which is easy to make the root system hang in the air and cannot grow down, seriously affecting the normal growth of seedlings. However, dry soil preparation can overcome the above problems. The rice seedling tray and the seedbed are closely attached; the root system absorbs water and nutrients quickly and grows down deeply, and the rice seedlings grow consistently and have good root packing. The basic process of seedling plate making by dry soil preparation and ventilation seedling raising is as follows: On the basis of plowing, freeze-thawing and rotary tillage, leveling should be carried out with a laser leveling instrument, and then mechanical ditching should be used for bordering. After the upturned soil is sunned for 2–3 days, water should be fed for manual leveling 1–2 times, and drainage and drying should be carried out to make the plate surface compacted. Two to three days before sowing, shoveling should be carried out again to fill the cracks, and the cracks should be fully compacted. The plate surface should meet the requirements of "compactness, flatness, smoothness, and straightness". The seedling plate specification should be a bed width of 1.50 m, furrow width of 0.25 m, and furrow depth of 0.20 m, so that irrigation and drainage can be separated, the inner and outer furrows can be matched, and irrigation, drainage, and dewatering can be achieved.

The obvious difference between rice seedling raising by potted seedling machine and carpet seedling raising is that fertilization in seedling fields needs to be strengthened and the concentration of soil nutrients increased. There are two main reasons: First, compared with carpet seedlings, rice seedlings transplanted by a potted seedling machine have a longer seedling age, and the root system of rice is deeply penetrated into the seedbed soil. The nutrients required for seedlings are mainly absorbed by the seedbed soil, so the nutrient concentration in the seedbed soil must be increased. Second, for the water layer management of rice seedling fields transplanted by a potted seedling machine, water layer can not be established for a long time, so dry raising is the main method. Otherwise, the root system of seedlings is easy to rise and cause root winding, which will seriously affect transplanting. Under the condition of dry raising, water control and fertilizer control are inevitable, so the fertilizer supply intensity must be increased to meet the nutrient supply required for the growth and development of seedlings, which requires enhanced fertilization of seedling fields, rather than top dressing alone (Top dressing is only an "early promotion", and a water layer must be established for top dressing; otherwise, it is easy to burn seedlings, but the establishment of a water layer will affect the effect of dry raising). Therefore, the fertilization of rice seedling fields transplanted by the potted seedling machine is the basis for raising strong seedlings, and it is impossible to raise strong seedlings if the seedling pool (field) is not strong.

The fertilization of potted seedlings is generally carried out in combination with soil preparation 30 days before seedling raising and after fine soil sieving. Inorganic fertilizer is generally used for fertilization, with a reference dosage of 60 kg of high-concentration nitrogen, phosphorus, and potassium compound fertilizer (the contents of available nutrients of nitrogen, phosphorus, and potassium are 15%, 15%, and 15%, respectively) and 30 kg of urea per mu of seedling field. After fertilizer spreading, it is necessary to carry out rotary tillage and burying fertilizer in time, plow ditches for bordering, and level the plate surface.

6. Precision sowing

The process and method of mechanical sowing of rice by potted seedling machine are obviously different from those by carpet seedling machine. The basic process of mechanical sowing of rice by potted seedling machine is as follows: placing the hard tray on the inlet of the sowing line → placing the subsoil → first compaction with iron balls → sowing → second compaction with iron balls → covering with soil → watering → centralized dark treatment for seedling emergence → placing the tray into the seedling field → seedling field management.

(1) Sowing.

Sowing amount: 50 g of hybrid rice and 70–75 g of japonica rice should be evenly sown in each tray. 6–7 seeds should be sown per hole for conventional japonica rice, 4–5 seeds per hole for hybrid japonica rice, and 3–4 seeds per hole for hybrid indica rice.

The mechanical sowing line should be used for quantitative sowing, and the sowing machine should be strictly commissioned to ensure optimal operation performance before sowing. The following aspects should be noted: ① The sowing machine should be adjusted to a horizontal state; otherwise, it will cause uneven sowing and uneven soil covering. ② The sowing amount should be adjusted, and the sowing groove wheels of different models should be selected according to different rice varieties (long grains or round grains). The sowing amount (average actual number of seeds per hole) should be adjusted in place before formal sowing. In adjusting the sowing amount, attention should be paid to the consistency between the sowing amount per hole and the sowing amount on the left and right sides of the seedling tray. ③ The amount of water sprayed should be adjusted to basically soak the soil. The amount of water sprayed should not be too large to avoid excessive water in the pot, resulting in hypoxia and rotten seeds, or flushing seeds and soil on the tray surface. Uncertain humidity easily causes a low seedling emergence rate and a low ejection rate of the pot during transplanting. ④ The amount of soil used should be adjusted, and the thickness of nutrient subsoil in the pot should be controlled to 2/3 of the hole depth, the covering soil should be controlled at the height of about 0.5 cm, and the thickness of the covering surface soil should not exceed the tray surface, so that no rice sprout is found.

(2) Dark treatment for seedling emergence.

The practice has proved that by superposing hard trays for dark treatment for seedling emergence, the temperature and humidity of each tray can be kept appropriate and consistent, the seedlings can emerge neatly and the hole rate is low, which not only solves the problem of difficult seedling emergence in production, but also reduces the waste tray rate and saves a large number of seedling trays, and reduces the requirements for seedbed quality through dark treatment technique, thus saving labor and cost. The dark treatment of rice planted by the potted seedling machine for seedling emergence is to superpose the seedling trays that have been sown in time for seedling emergence in a place directly exposed to sunlight outdoors. Attention should be paid to the upper and lower layers of seedling trays being vertically crossed during superposition to ensure that the holes of the upper seedling tray are placed on the groove of the lower seedling tray. The number of superposed seedling trays is 20, and there should be a certain gap of about 30 cm between each stack of seedling trays. To ensure that the temperature and humidity of each seedling tray under dark treatment are as consistent as possible, the bottom of the lowest tray of each stack should be padded with thermal insulation materials or an empty seedling tray for support, and a seedling tray without sowing but with soil and water should be placed on the top of each stack. After the superposition of the seedling trays is completed, the seedling trays should be covered with black plastic cloth in time and compacted with fine soil. After that, the plastic cloth should be lifted regularly everyday to check the surroundings of the stacked seedling trays. In the case of water shortage in the holes around the seedling tray, watering should be conducted with a watering can in time. After dark treatment for 3–4 days, when the incomplete rice leaves are grown, the plastic cloth can be removed and the tray can be placed.

(3) Sequential laying of trays.

The bed surface should be paved with a root-cutting mesh before trays are placed. The so-called root-cutting mesh is fine-mesh nylon gauze (mesh area < 0.25 cm^2). The fine-mesh nylon gauze should be paved on the bed surface before the tray placement to prevent the bottom of the tray from sticking to the soil, which is not conducive to seedling pulling. The tray placement is to directly place the seedling trays upon dark treatment on the bed side by side along the length direction of the tray, and the trays are closely laid between each other. The seedling trays should be closely attached to the bed surface with no space. The trays on the seedling plate should be placed flat and neat. After the tray is placed, the seedlings should be immediately irrigated once for seedling emergence to achieve quick irrigation and rapid drainage. Each seedling plate has two rows of trays laid horizontally in a tight and neat manner. The flashes between the trays should be overlapped, and the bottom of the trays should be closely attached to the bed surface.

(4) Covering with film and straw.

When the humidity of the soil in the pot reaches 75%–85% of the saturated water content, it is necessary to cover the agricultural film flatly and evenly cover the film surface with straw. The amount of dry straw covered is 0.6 kg/m^2. It will increase the covering amount appropriately in sunny weather or reduce it in gloomy weather. The film should be basically invisible. The water discharge gap should be opened around the seedling field to avoid water accumulation due to precipitation during the seedling emergence stage, which will result in rotten buds.

7. Seedling stage management

(1) Prevention of covering damage by wind and animals.

After sowing, it is necessary to frequently observe whether the straw covered is damaged by livestock and wind. Once damaged, it must be repaired immediately.

(2) Timely seedling hardening.

After full emergence, the film for seedling hardening shall be revealed when the second complete leaf comes out by 1.0–3.0 cm. Requirements for film uncovering: The film can be uncovered in the evening on sunny days, in the morning on cloudy days, before light rain and after heavy rain.

(3) Scientific water management.

Replenish sufficient water once on the day when the film is uncovered, and replenish water as the film is uncovered. Then, observe the leaf roll of the seedlings. Do not replenish water if the leaves are not rolled, and replenish water to the place where the leaves are rolled.

(4) Disease and pest control.

The method is the same as that of carpet seedling raising.

8. Preparation before transplanting

(1) Top dressing of seedlings.

For potted seedlings, on the basis of strengthening seedbed fertilization in seedling fields, attention should also be paid to the top dressing of seedlings as early as possible to prevent the seedlings from yellowing and ensure long and thick rice in the 4-leaf stage with tillering. The main measures include applying “weaning fertilizer” early, providing a timely nutrition supplement, and facilitating the transformation of rice seedlings from heterotrophic to autotrophic. “Weaning fertilizer” is applied in the 2-leaf stage, with 4 g of compound fertilizer per tray. The total available nutrient content of nitrogen, phosphorus, and potassium in compound fertilizer is ⩾45%, accounting for 15%, 15%, and 15%, respectively. After fertilization, clean water is gently sprinkled with a watering can to prevent seedlings from burning. The main target of transplanting in the 4-leaf stage is to improve the ability to resist transplanting injury and the rooting ability after transplanting. The key is to improve the carbon and nitrogen nutrient content of seedlings to control water, strengthen roots, and improve seedlings. The main measure is to apply transplanting fertilizer and pay attention to water control. Transplanting fertilizer should be applied 2–3 days before transplanting, and about 5 g of compound fertilizer should be used for each tray.

(2) Timely chemical control to prevent vigorous growth of seedlings.

Potted seedlings should be raised with the seedling hardening agent in nutrient soil. As it contains paclobutrazol, it has a certain control effect on the height of seedlings. However, when the seedling age exceeds 20 days, the effect of the seedling hardening agent on controlling height gradually disappears, and the seedlings are obviously higher. Therefore, when the age of potted seedlings reaches 30 days, paclobutrazol must be applied separately for chemical control to prevent the seedlings from growing vigorously, and the height of seedlings must be controlled not to exceed 20 cm to adapt to machine transplanting. According to the test results, within a certain range of seedling age, the chemical control of seedlings can be carried out in 2 rounds. For the first round, 4–5 g of 15% paclobutrazol powder can be used for every 100 seedling trays at the 2-leaf stage, and for the second round, 5–6 g of 15% paclobutrazol powder can be used for every 100 seedling trays at the 4-leaf stage. The paclobutrazol powder should be sprayed with water, and the spray should be uniform and careful. If the seedlings are at an old foliar age or the seedling age is old due to a delayed machine transplanting stage during the spray of paclobutrazol power, it is necessary to increase the dosage appropriately.

(3) Pest control.

It is necessary to pay close attention to the occurrence of underground pests, plant hoppers, *Stenchaetothrips biformis* (Bagnall), rice bakanae, seedling blasts, and other diseases and pests in the seedling stage. After the tray is placed, chemicals should be applied once every 2–3 days to prevent and control *Laodelphax striatellus* (Fallen), which means spraying 80 mL of 48% chlorpyrifos and 20 mL of chlorantraniliprole per mu evenly before evening. Pesticides should be sprayed before transplanting to achieve transplanting with pesticides.

Section III Field Transplanting Technique for Machine-transplanted Rice

I. Introduction to High-performance Rice Transplanter

1. Introduction to rice transplanter

(1) Working principle and classification of rice transplanter.

At present, the working principles of mature and widely used rice transplanters at home and abroad are generally the same. The engine transmits the power to the seedling transplanting mechanism and the seedling conveying mechanism respectively. With the cooperation of the two mechanisms, the seedling needle of the seedling transplanting mechanism is inserted into the seedling block to grab the seedlings, take them out, and move them down. When they are moved to the set seedling transplanting depth, the transplanting fork in the seedling transplanting mechanism presses the seedlings down from the seedling needle to complete a seedling transplanting process. Meanwhile, the relative position of the traveling wheel and the machine body, as well as the relative position of the floating plate and the seedling needle, are controlled through the floating plate and the hydraulic system, so that the seedling transplanting depth is basically consistent.

Rice transplanters are usually classified by the operation mode and transplanting speed. According to the operation mode, the rice transplanter can be divided into walking-type rice transplanter and riding-type rice transplanter. According to the transplanting speed, it can be divided into ordinary rice transplanter and high-speed rice transplanter. Currently, all walking-type rice transplanters are ordinary rice transplanters; riding-type rice transplanters include ordinary rice transplanters and high-speed rice transplanters. According to the number of rows transplanted, the walking-type rice transplanter can transplant 2, 4, or 6 rows, while the riding-type rice transplanter can transplant 4, 5, 6, 8, or 10 rows. According to seedlings transplanted, there are carpet seedling transplanters and potted seedling transplanters. Carpet seedling transplanters are widely used in Jiangsu. The structure of the potted seedling transplanter is complicated, and a special rice seedling tray is required, which is costly to use.

(2) Main technical characteristics of rice transplanter.

First, the basic seedlings, transplanting depth, plant spacing, and other indicators can be quantitatively adjusted. The basic seedlings transplanted by the rice transplanter are determined by the number of holes (density) transplanted per mu and the number of plants per hole. According to the requirements for rice population quality, cultivation, row expansion, and seedling reduction, the row spacing of seedlings of rice transplanters is fixed at 30 cm, and the plant spacing can be adjusted in multiple levels or steplessly to reach the transplanting density of 10,000–20,000 holes per mu. The area of small seedling blocks taken (number of seedlings per hole) is adjusted by adjusting the transverse movement handle (multi-level or stepless) and the longitudinal seedling conveying adjustment handle (multi-level) to achieve a suitable basic seedling. At the same time, the insertion depth can also be conveniently and accurately adjusted through the handle, which can fully meet the agronomic technical requirements.

Second, the rice transplanter has a hydraulic profiling system to improve the stability of paddy field operation. It can continuously adjust the machine status with the fluctuation of the field surface and hard subsoil to keep the machine balanced and insertion depth consistent. With the difference of soil hardness on the soil surface due to the method of field preparation, certain ground pressure of the board is maintained to avoid strong dammed mud and drainage, which will affect the transplanted seedlings.

Third, the degree of mechatronics is high, and the operation is flexible. The high-performance rice transplanter has world-class mechanical technology and a high degree of automation control and mechatronics, which fully ensures the reliability, adaptability, and operation flexibility of the machine.

Fourth, it is highly efficient, labor-saving, and cost-effective. The maximum operation efficiency of the walking-type rice transplanter is up to 4 mu/hour, and that of the riding-type high-speed rice transplanter is 7 mu/hour. Under normal working conditions, the operation efficiency of the walking-type rice transplanter is generally 2.5 mu/hour, and that of the riding-type high-speed rice transplanter is 5 mu/hour, which is much higher than the efficiency of manual transplanting.

2. Carpet seedling transplanter

There are two types of carpet seedling transplanters: walking-type transplanters and high-speed transplanters.

The walking-type transplanter is a kind of cheap transplanter that is suitable for the current rural natural and economic conditions in China. The so-called "walking type" is the description of holding the handle of the machine and walking forward to control the machine. The machine is simple in structure, light in weight, easy to operate, safe and reliable in use, and easy to control the transplanting quality. Walking-type transplanters have 2-, 4-, and 6-row types depending on the number of rows. Generally, 4-row transplanters are used, and 6-row transplanters are more difficult to operate with higher efficiency. There are many types produced by domestic, joint ventures, and sole proprietorship enterprises in the market, such as Fulaiwei, TYM, ISEKI, Kubota, and ASIA, with similar structures. Since the walking-type rice transplanter is not widely used at present, it will not be detailed in this book. Here are the common high-speed rice transplanters in the market.

(1) Model and features.

The high-speed rice transplanter is a high-tech model that has the advantages of comfort and high efficiency compared with the walking type, and follows the trend of riding. The high-speed rice transplanter has 4, 5, 6, 8, and 10 rows and other types. The more rows there are, the higher the efficiency is. However, the machine is heavy and expensive. Generally, a 6-row type is used, and the engine power is between 7 and 12 horsepower, so it is more cost-effective. Now, the P600 high-speed rice transplanter is taken as an example (Table 3-3), which is a reference for the use of other models.

Table 3-3 Main Technical Specifications of TYM P600 High-speed Rice Transplanter

Model				P600
Type				Riding, 6-row
Dimensions		Length (mm)		2,960(3,020 at work)
	Width (mm)		2,010(2,870 at work)	
	Height (mm)		1,450(1,540 at work)	
Weight (kg)				550
Engine	Model			FE290G-7SX
	Total displacement (mL)			286
	Power/Speed (ps/rpm)			7.0/1,800 (8.06 max.)
	Fuel consumption (g/ps·hr)			300
	Starting mode			Engine
Driving part	Wheel Outer diameter (mm)		Type	Rubber-coated vane wheel
			620 (front wheel)/ 850 (rear wheel)	
	Number of gears			Forward: 3; backward: 1 (5 gears for auxiliary transmission)
Rice transplanting part	Type of rice transplanting arm			Rotary
	Seedling conveying mode/Seedling box capacity			Belt/2.2 pieces
	Number of rows/Row spacing (cm)			6/30
	Plant spacing (cm)/Number of holes per 3.3 m^2			14.5,12.8,11.3/75,85,95
	Transplanting depth (cm, adjustable)			0–4.6
	Transverse seedling collection amount (times)/ Longitudinal seedling collection amount (mm)			18, 20, 24 (3 gears adjustable) /8–17
Efficiency				4.7–7.5
	Transplanting speed (m/s)			0.15–1.2

continued

Model		P600
Type		Riding, 6-row
Conditions of seedlings	Types of seedlings	Seedling raising by seedling tray
	Seedling height (cm)	Young seedlings: 10–18, medium seedlings: 15–25 (30 max.)
	Foliar age (leaf)	2–5
Field conditions	Tilling depth (cm)	8–20
	Water depth (cm)	1–4

1) Features of TYM P600 high-speed rice transplanter.

① The rotary rice transplanting arm features high speed (transplanting twice in one circle).

② Adopt the hydraulic profiling device with automatic adjustment.

③ Use the monitor to understand the seedling supply time and the status of the rice transplanting arm. When the rice transplanting clutch is in the "Rice Transplanting" position and the "No Rice Transplanting" position, there are indicator lamps for the prompt. When rice seedlings need to be supplied, an alarm sound will be given.

④ Four-wheel drive is adopted, which is very convenient for entering and leaving the field and crossing the ridge and ditch.

⑤ 5-segment hydraulic sensor with high sensitivity can automatically adjust to adapt to the hardness of the plot.

⑥ The transplanting depth and plant spacing can be easily adjusted.

⑦ The hydraulic power-assisted steering wheel is easy and convenient to operate.

2) Name of each part of TYM P600 high-speed rice transplanter (Fig. 3-3).

3) Functions of each part of TYM P600 high-speed rice transplanter.

① Switch function.

Key switch (Fig. 3-4): OFF (the engine stops); ON (the engine works).

Horn switch (Fig. 3-5): Press it to sound the horn.

Automatic scribing switch and indicator (Fig. 3-6): When the switch is pressed (ON), the automatic scriber can be started (the indicator lamp is on).

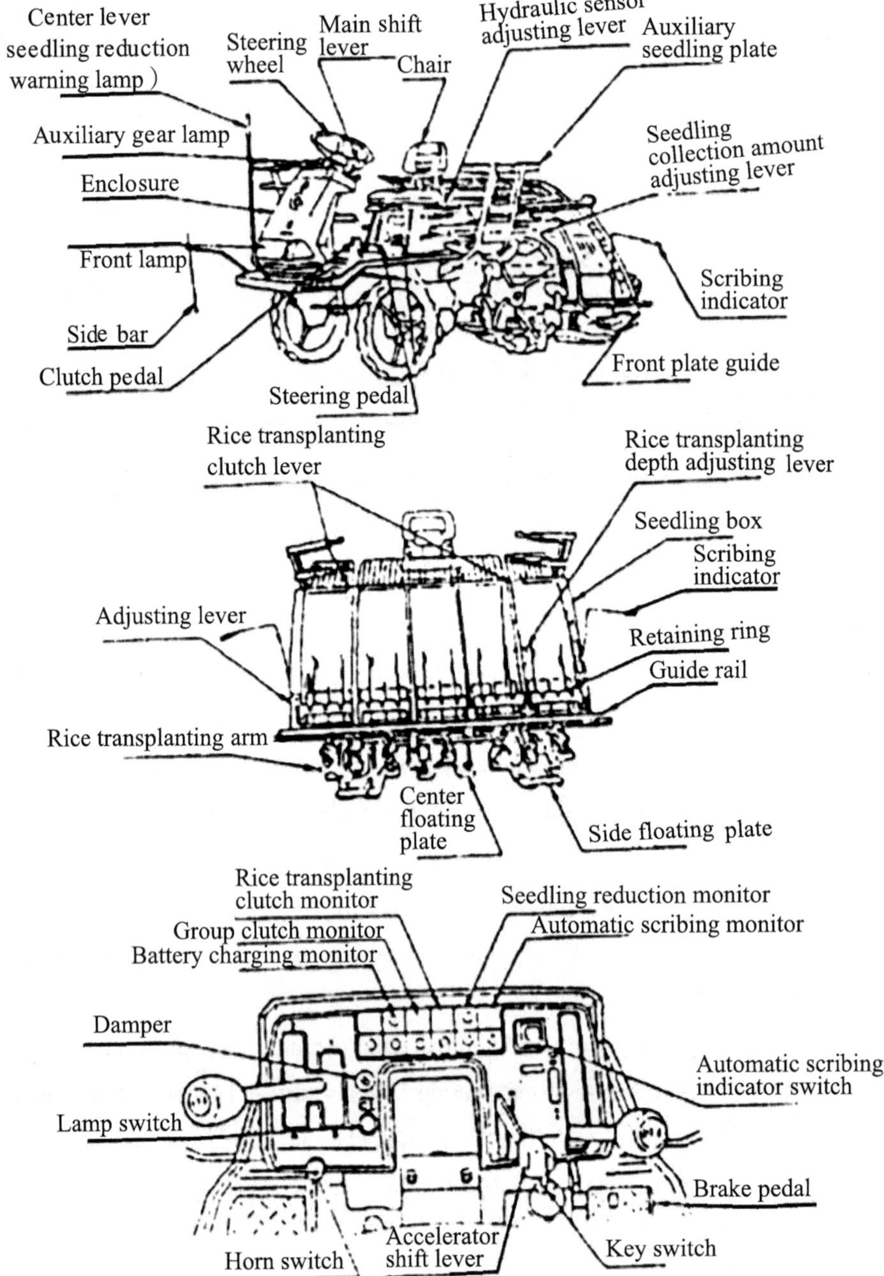

Fig. 3-3 Name of Each Part of TYM P600 High-speed Rice Transplanter

Press the switch again, and the automatic scriber will stop working (the indicator lamp will go out).

Front lamp switch (Fig. 3-7): ON (the front lamp is on); OFF (the front lamp is off).

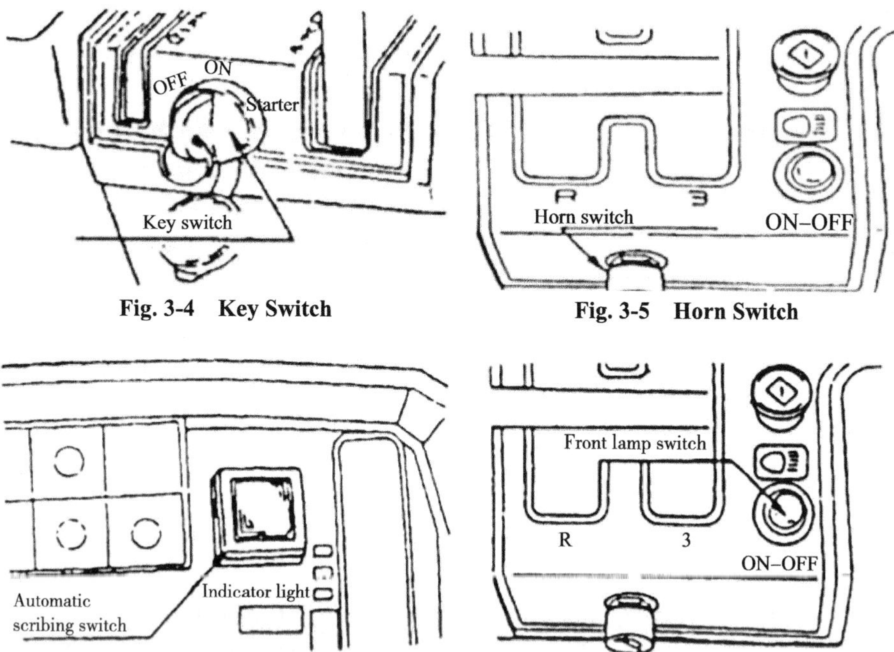

Fig. 3-4 Key Switch

Fig. 3-5 Horn Switch

Fig. 3-6 Automatic Scribing Switch and Indicator

Fig. 3-7 Front Lamp Switch

② Functions of monitors.

Group clutch monitor (Fig. 3-8): When the group clutch lever is in the "connected" position, the indicator lamp is on.

Rice transplanting clutch monitor (Fig. 3-8): When the rice transplanting clutch lever is in the "transplanting" position, the indicator lamp is on.

Seedling reduction monitor (Fig. 3-9): This indicator lamp will be on when at least a certain amount of seedlings are in the seedling box.

Battery charging monitor (Fig. 3-9): When the battery voltage becomes low, this indicator lamp will be on to prompt charging as soon as possible.

Center lever (Fig. 3-10): It is a marker lever for the driver to drive in a straight line, and there is an indicator lamp at the top of the center lever. When the number of seedlings in the seedling box is less than a certain amount, the indicator lamp will be on, prompting to add seedlings as soon as possible.

Scriber monitor (Fig. 3-11): The indicator lamp is on, and the scriber enters the working status.

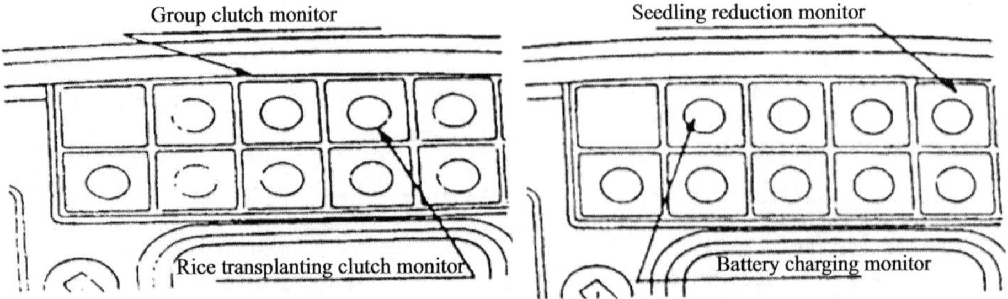

Fig. 3-8 Rice Transplanting Clutch Monitor **Fig. 3-9 Battery Charging Monitor**

Fig. 3-10 Center Lever **Fig. 3-11 Scriber Monitor**

③ Functions of each operating device.

Accelerator shift lever (Fig. 3-12): When the lever is pulled up, the engine speed increases; when the lever is pushed down, the engine speed decreases.

Accelerator pedal (Fig. 3-13): When the pedal is depressed, the engine speed increases.

Damper (Fig. 3-14): When the engine is cold, pull up the damper handle to start the engine. After starting, the damper should be pushed back.

Main shift lever (Fig. 3-15): The vehicle speed can be changed, and there are 3 gears for driving (1st and 2nd gears for rice transplanting, 3rd gear for driving), 1st gear, PTO shaft and neutral gear for backward.

Auxiliary shift lever (Fig. 3-16): When used with the main shift lever, the vehicle speed can be changed. Push the lever outward to accelerate and pull it inward to decelerate.

Rice transplanting clutch lever (Fig. 3-17): This lever is used to control the lifting and lowering of the rice transplanting device and the power transmission of the rice transplanting device.

Steering wheel angle adjusting lever (Fig. 3-18): It is a steering wheel locking device. When the device is turned on, the position of the steering wheel can be adjusted. After adjustment, lock it with the adjusting lever.

Plant spacing changing lever (Fig. 3-19): It can be adjusted up and down to change the plant spacing during rice transplanting. Note: This lever can only be adjusted when the main shift lever is in the "PTO" position and the engine is running at a low speed.

Plant spacing auxiliary shift jack (Fig. 3-20): Select the standby jack for the most suitable plant spacing. Note: When it cannot be changed, please move the rice transplanting arm first.

Clutch pedal (Fig. 3-21): It is used to connect or disconnect the main clutch. When depressed, the clutch is cut off; when depressed again, the clutch will be reconnected.

Brake pedals (Fig. 3-22): When the left pedal is depressed, it is connected with the left rear wheel; when the right pedal is depressed, it is connected with the right rear wheel. When turning, step on the pedal on one side of the turn. Connect the left and right brakes when not in the field, especially when driving.

Brake lever (Fig. 3-23): It can stop the rice transplanter.

Steering pedal (differential locking pedal) (Fig. 3-24): It is used when the rear wheels of the rice transplanter slip. Do not step on the pedal when turning the steering wheel.

Hydraulic sensor adjusting lever (Fig. 3-25): Change the sensitivity of the hydraulic sensor according to the hardness of the plot. When the lever is in the "Fixed" position, the rice transplanting device will not descend.

Rice transplanting depth adjusting lever (Fig. 3-26): It is used to adjust the rice transplanting depth. The depth is shallow when it is fixed on the top and deep when fixed on the bottom.

Group rice transplanting clutch lever (Fig. 3-27): There are three groups, which control the operation of two rows of rice transplanting arms in each group respectively. It is used to adjust the number of rows of seedlings transplanted according to the field conditions.

Longitudinal seedling collection amount adjusting lever (Fig. 3-28): It is used to adjust the amount of seedlings collected longitudinally, and the seedling collection amount is the smallest when the lever is moved to the right side.

Transverse conveying capacity adjusting lever (Fig. 3-29): It is used to adjust the transverse conveying capacity when the seedling box moves to both ends.

Release lever for seedling pressing (Fig. 3-30): When lifting out the seedlings, please place the lever in the "withdrawal" position.

Seedling retaining ring (Fig. 3-31): It is used to retain the seedlings in the row when no rice transplanting is required in the side row. The transplanting arm cannot catch the seedlings, and the rice transplanting is stopped. When stopping the two rows of rice transplanting arms, use the group rice transplanting clutch.

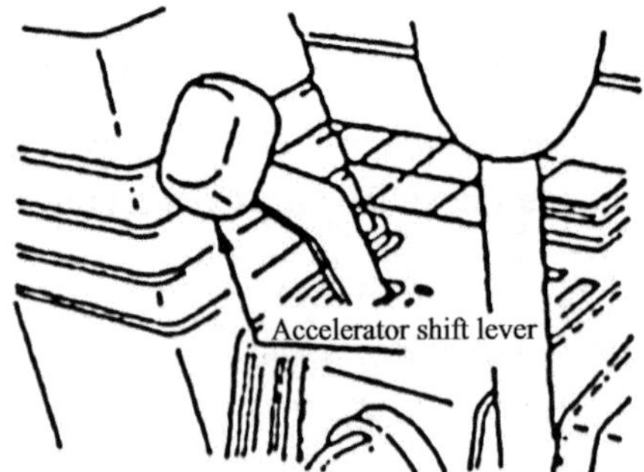

Fig. 3-12 Accelerator Shift Lever

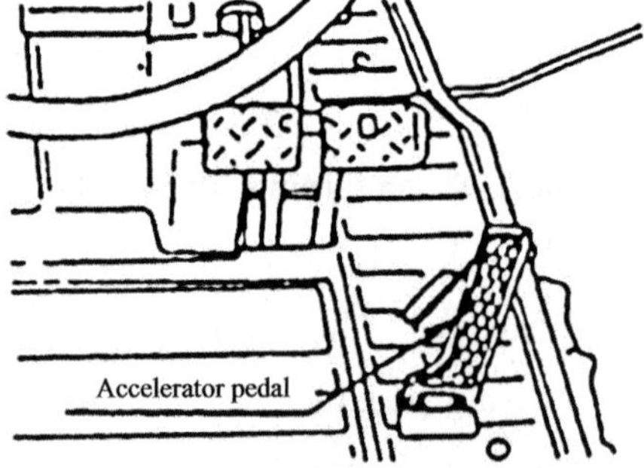

Fig. 3-13 Accelerator Pedal

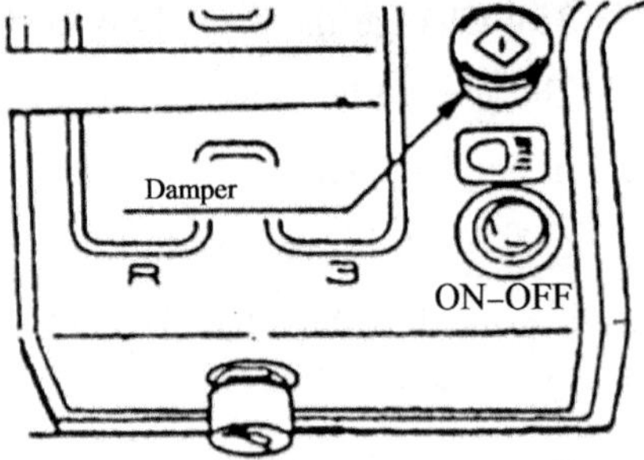

Fig. 3-14 Damper

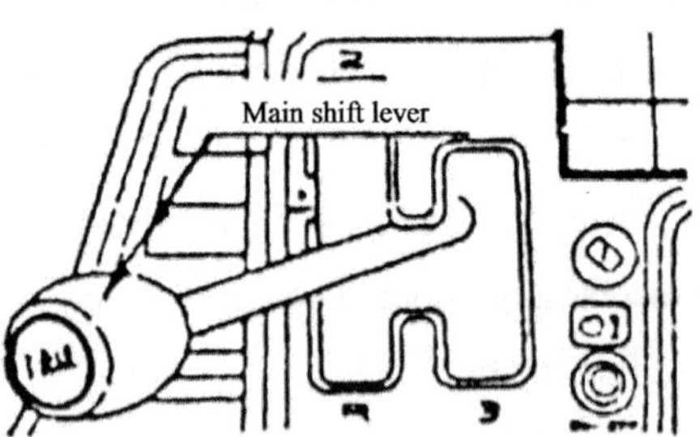

Fig. 3-15 Main Shift Lever

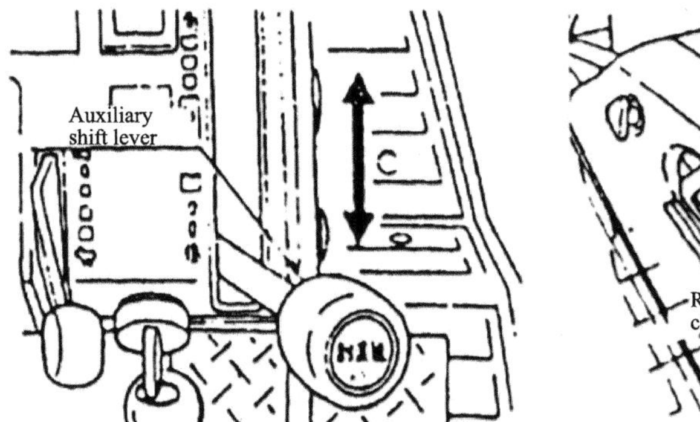

Fig. 3-16 Auxiliary Shift Lever

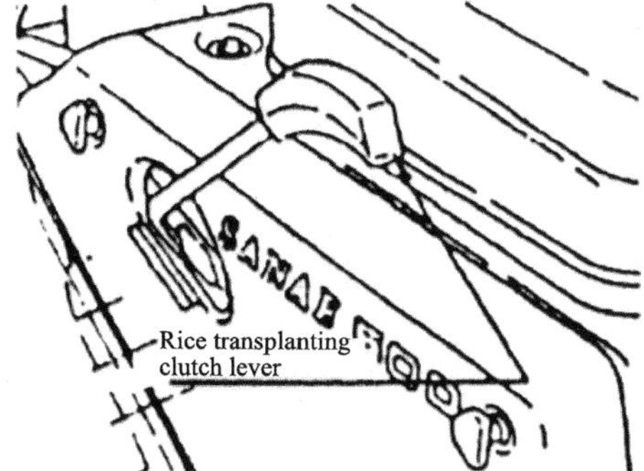

Fig. 3-17 Rice Transplanting Clutch Lever

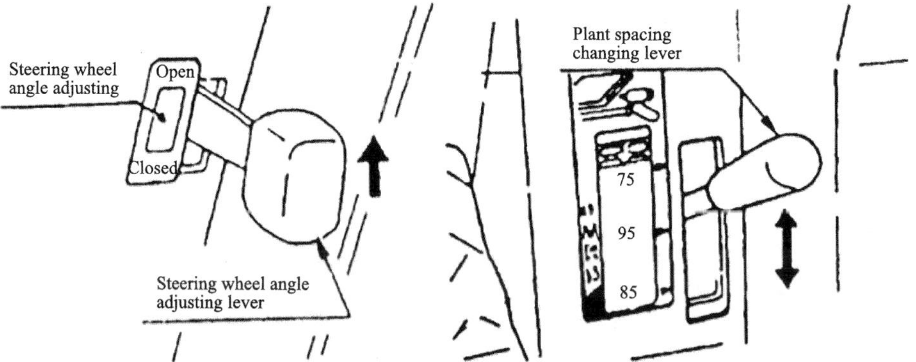

Fig. 3-18 Steering Wheel Angle Adjusting Lever

Fig. 3-19 Plant Spacing Changing Lever

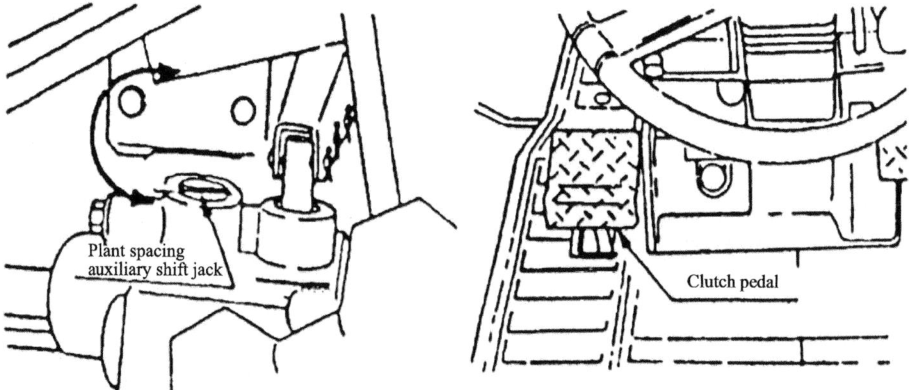

Fig. 3-20 Plant Spacing Auxiliary Shift Jack

Fig. 3-21 Clutch Pedal

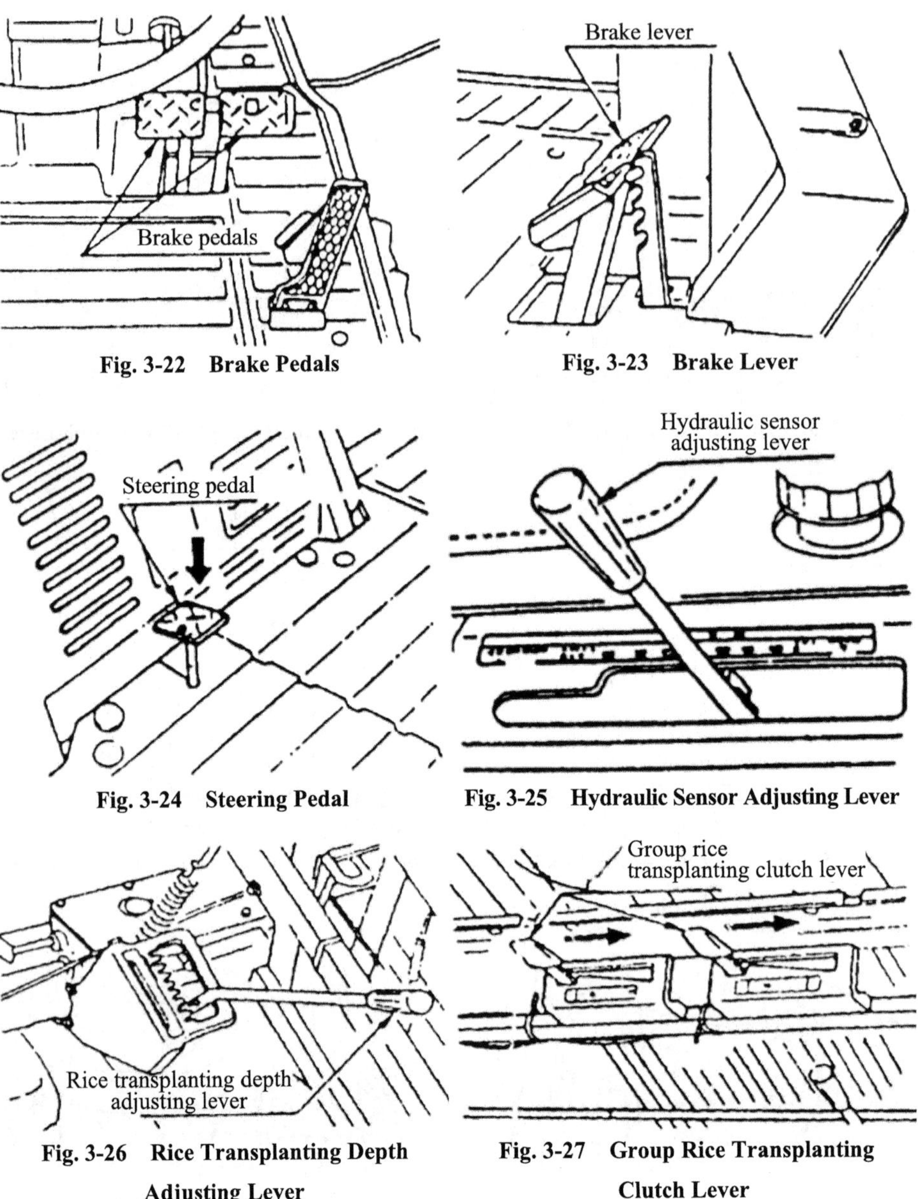

Fig. 3-22 Brake Pedals

Fig. 3-23 Brake Lever

Fig. 3-24 Steering Pedal

Fig. 3-25 Hydraulic Sensor Adjusting Lever

Fig. 3-26 Rice Transplanting Depth Adjusting Lever

Fig. 3-27 Group Rice Transplanting Clutch Lever

(2) Operating technique.

1) Startup of engine.

Turn on the fuel switch: Put the fuel switch in the "ON" position (downward) (Fig. 3-32 and Fig. 3-33).

Clutch lever: Place the rice transplanting clutch lever in the "Fixed" position.

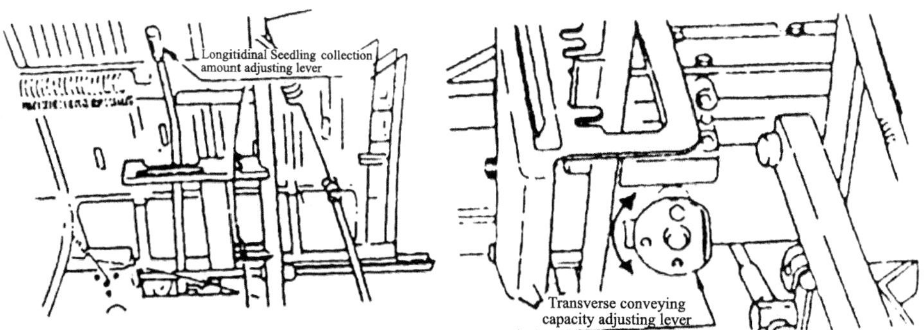

Fig. 3-28 Longitudinal Seedling Collection Amount Adjusting Lever

Fig. 3-29 Transverse Conveying Capacity Adjusting Lever

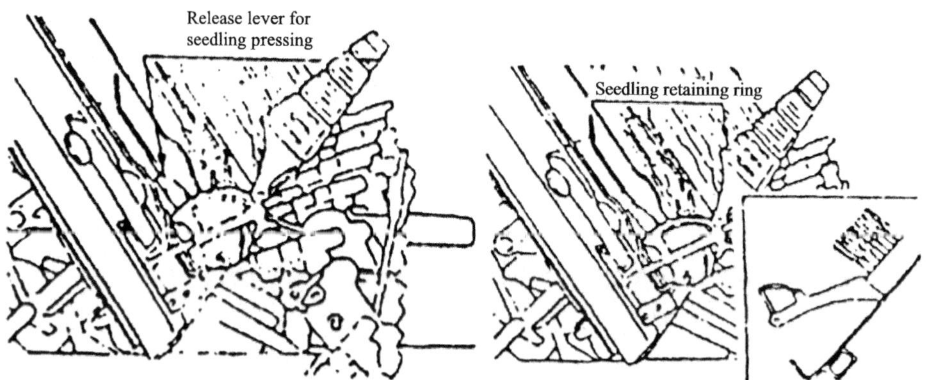

Fig. 3-30 Release Lever for Seedling Pressing

Fig. 3-31 Seedling Retaining Ring

Manual throttle: Put the manual throttle in the "Middle" position.

Pull up the damper: Pull up the damper handle.

Clutch: Depress the clutch pedal and disconnect the clutch (Fig. 3-34).

Ignition: Turn the key on the key switch to the "Ignition" position (Fig. 3-35).

Note: If it cannot be started within 10 seconds, please turn the key back to "OFF" and wait for 30 seconds before starting.

Push the damper: After the engine is started, push the damper in.

2) Shutdown of engine.

Push the manual throttle outward to run the engine at a low speed.

Then, turn the key on the key switch to the "OFF" position to stop the engine. As the battery will discharge when the key switch is in the "ON" position, please

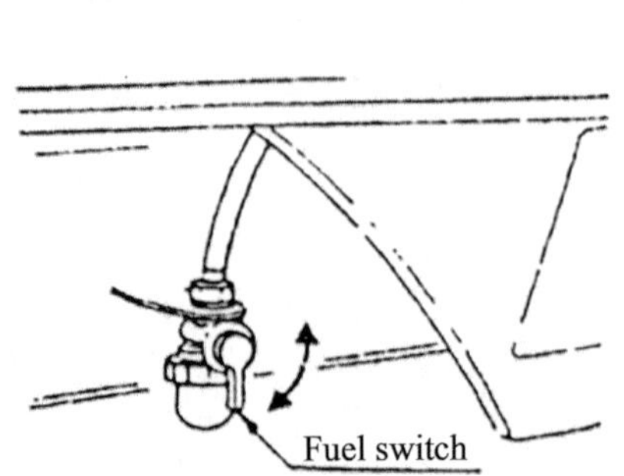

Fig. 3-32 Position of Fuel Switch

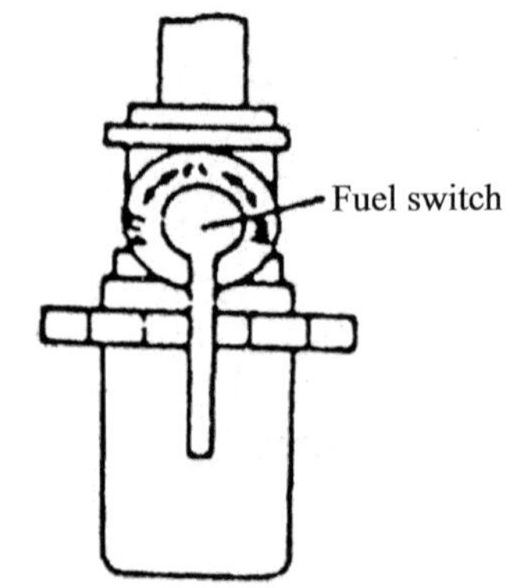

Fig. 3-33 Enlarged View of Fuel Switch

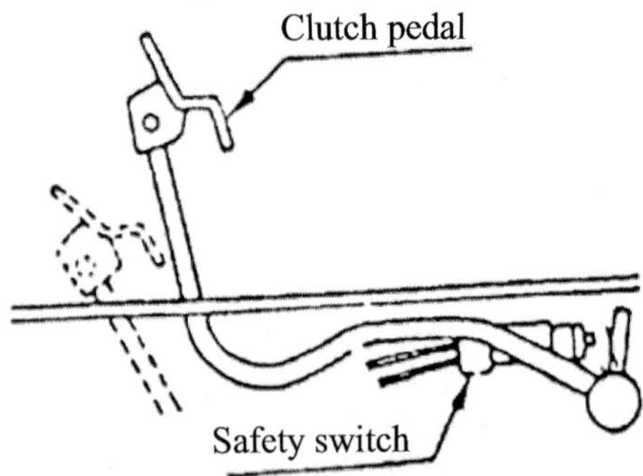

Fig. 3-34 Depressing the Clutch Pedal

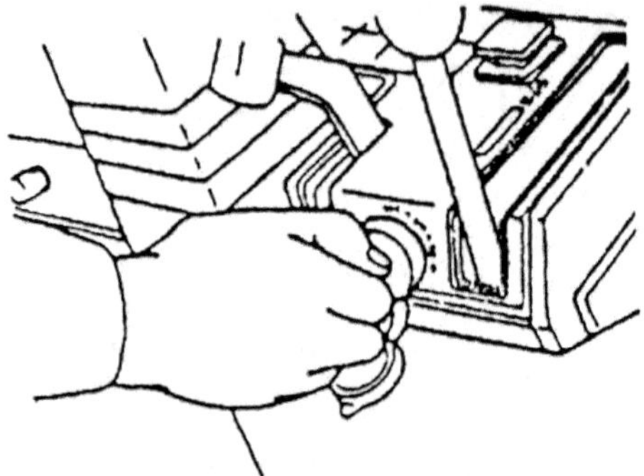

Fig. 3-35 Ignition

develop the habit of removing the key after the engine stops.

3) Method of startup.

First, depress the clutch pedal. Then, place the main and auxiliary shift levers in proper positions. Finally, slowly release the clutch pedal (Fig. 3-36).

Note: It is dangerous to release the clutch pedal quickly.

4) Method of shutdown.

First, push the manual throttle of the shift lever outward to slow down the engine. Then, turn the key switch to the "OFF" position to stop the engine. Finally, depress the clutch pedal and brake pedal at the same time (Fig. 3-37).

5) Road-based driving.

Step 1: Hang the scribing indicator (Fig. 3-38).

Step 2: Pull up the side indicator (Fig. 3-39).

Step 3: Remove the clamp of the front guide and adjust the position of the

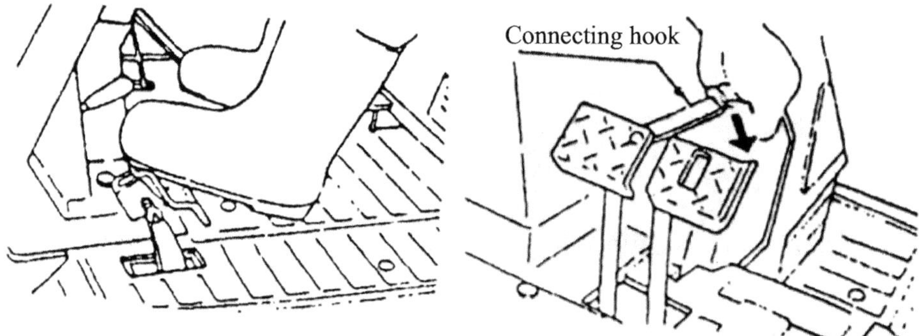

Fig. 3-36 Schematic Diagram of Startup **Fig. 3-37 Schematic Diagram of Shutdown**

guide (Fig. 3-40).

Step 4: Use the connecting hook to fasten the pedals of the left and right brakes.

Step 5: Place the rice transplanting clutch lever in the "up" position and lift the rice transplanting device completely.

Step 6: Place the hydraulic sensor adjusting lever at the position of "Lowering and fixing the rice transplanting device" (Fig. 3-41).

Step 7: Place the main shift lever at the "PTO" position, place the rice transplanting clutch lever at the "connected" position, move the seedling box to the center of the machine, and place the rice transplanting clutch lever at the "OFF" position. Select the main and auxiliary shift levers according to the actual situation.

6) Method of entering and leaving the field.

Step 1: Lift the rice transplanting device completely.

Step 2: Place the main shift lever on "1" and the auxiliary shift lever on "low speed", and then slowly drive into the field (Fig. 3-42).

Note: ① Be sure to use a springboard. ② Connect the pedals of the left and right brakes. ③ When entering the field, do not load seedlings or goods in the auxiliary seedling plate and seedling box. ④ When entering the field, the rice transplanter should be perpendicular to the field edge (Fig. 3-43).

7) Method of leaving the field.

Step 1: Lift the rice transplanting device completely.

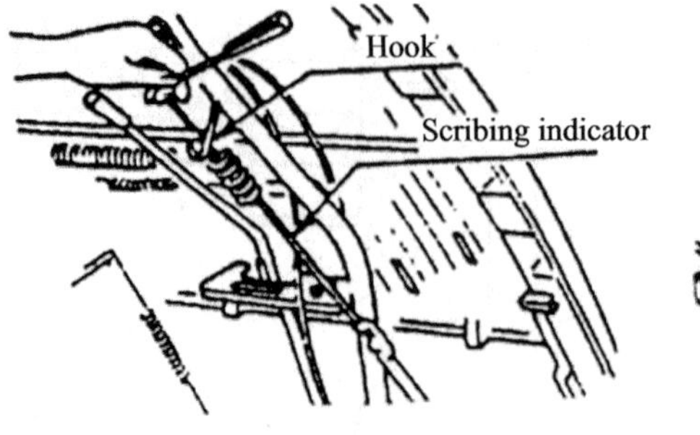

Fig. 3-38 Scribing Indicator

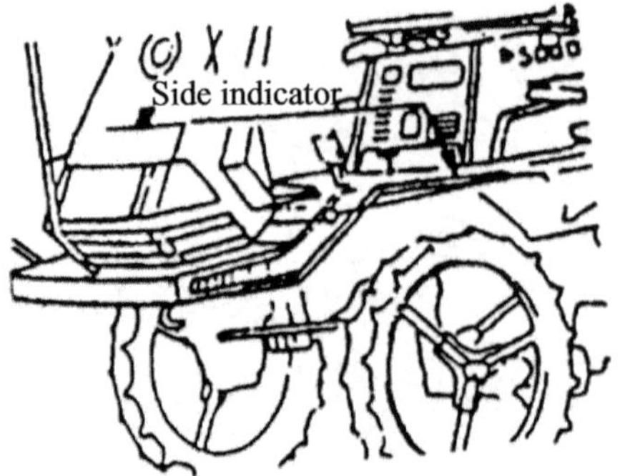

Fig. 3-39 Side Indicator

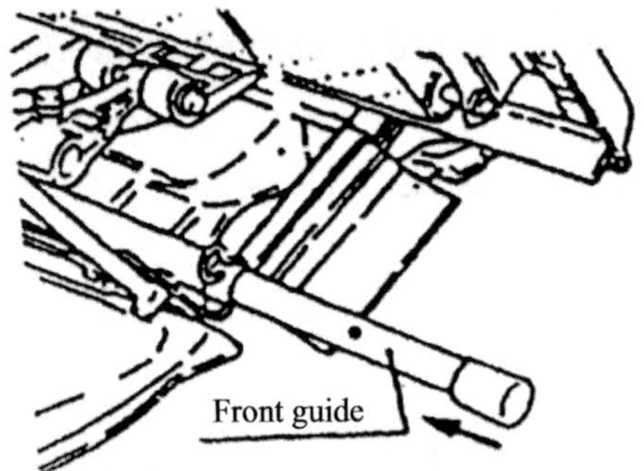

Fig. 3-40 Adjusting the Guide

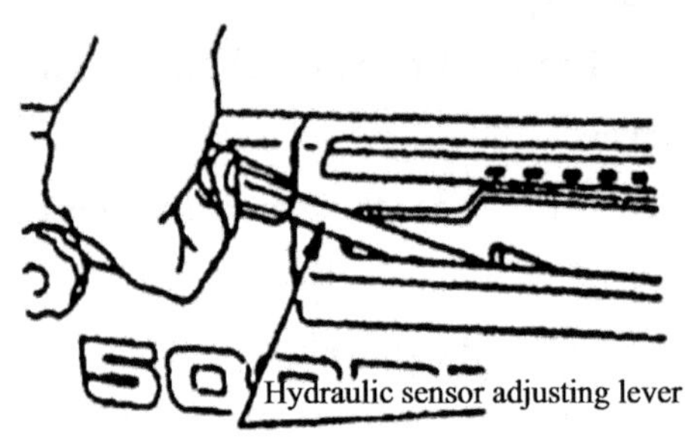

Fig. 3-41 Hydraulic Sensor Adjusting Lever

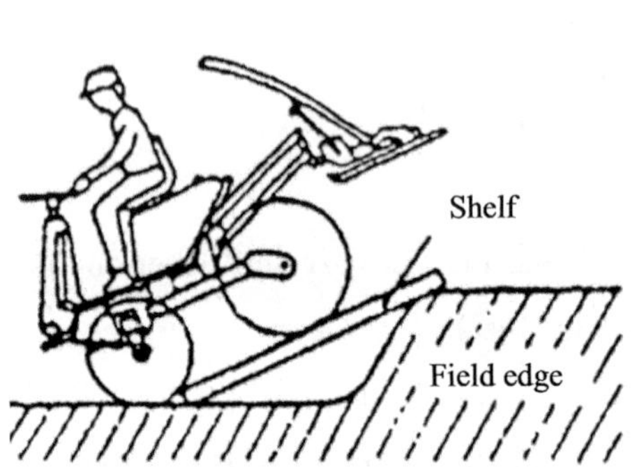

Fig. 3-42 Method of Entering the Field

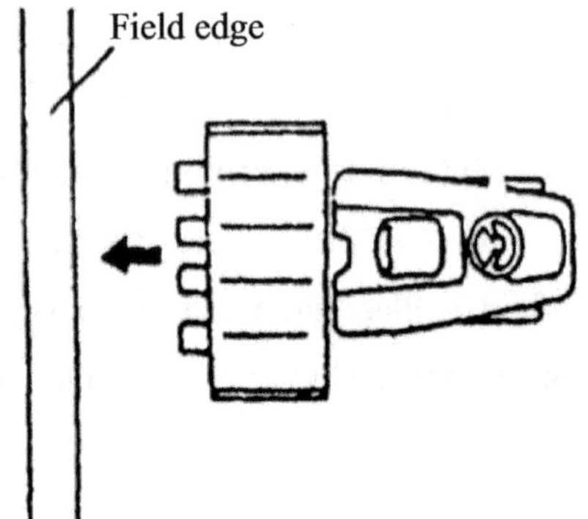

Fig. 3-43 Perpendicular to the Field Edge

Step 2: Place the main shift lever on "backward" and the auxiliary shift lever on "low speed", and then slowly come out of the field.

8) Adjustment method of plant spacing (Fig. 3-44).

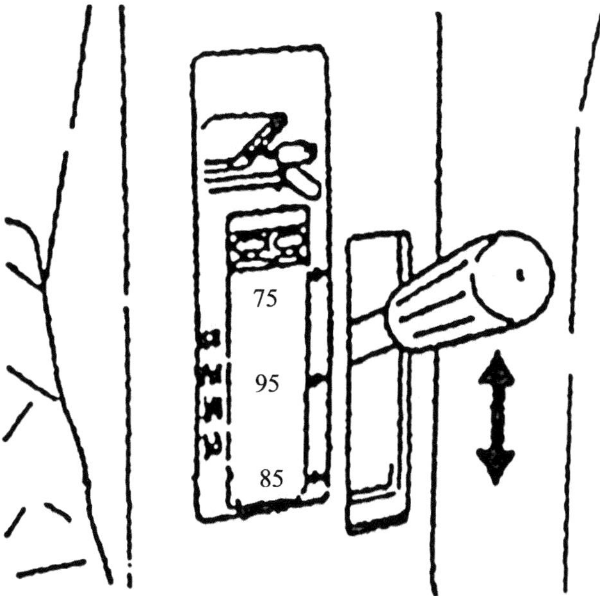

Fig. 3-44 Adjustment Method of Plant Spacing

Step 1: Completely lift the rice transplanting device and place the hydraulic sensor adjusting lever at the position of "Lowering and fixing the rice transplanting device".

Step 2: Run the engine at a low speed, and place the main shift lever in the "PTO" position.

Step 3: Place the rice transplanting clutch lever in the "Connected" position and rotate the rice transplanting arm.

Step 4: Move the plant spacing adjusting the lever up and down to find a suitable position.

Step 5: After the lever is adjusted, check whether the rice transplanting arm can rotate.

9) Adjustment method of transverse seedling collections.

The number of transverse seedling collections is adjusted according to the types of seedlings, as shown in Table 3-4 below. At the time of leaving the factory, the transverse seedling collections of the rice transplanter are set 24 times (Fig. 3-45 and Fig. 3-46).

Table 3-4 Number of Transverse Seedling Collections for Different Seedling Sizes

Types of seedlings	Number of transverse seedling collections
Small seedlings	24 times
Medium seedlings	20 times
Medium/Large seedlings	18 times

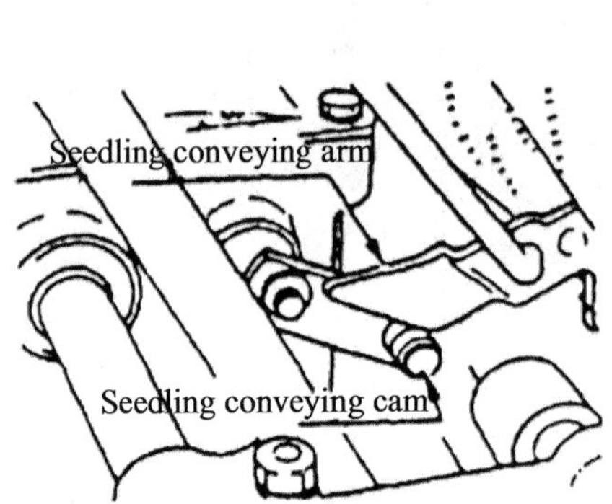

Fig. 3-45 Position of Seedling Conveying Cam during Transverse Seedling Collection Adjustment

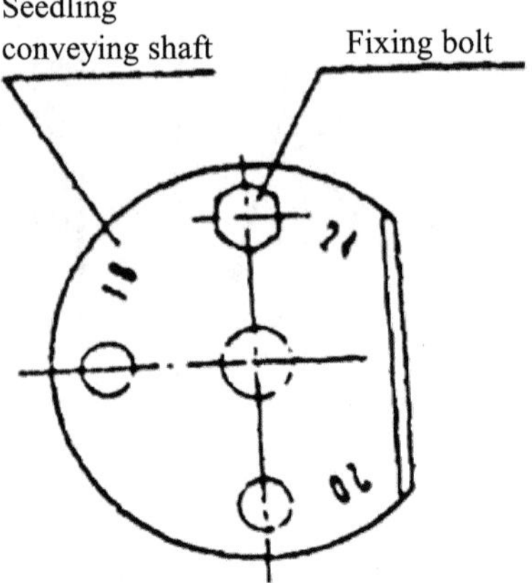

Fig. 3-46 Structure Diagram of Transverse Seedling Collection Adjustment Method

Step 1: Completely lift the rice transplanting device and place the hydraulic sensor adjusting lever at the position of "Lowering and fixing the rice transplanting device".

Step 2: Run the engine at a low speed, and place the main shift lever in the "PTO" position.

Step 3: Place the rice transplanting clutch lever at the "Connected" position, move the seedling box to any end on the left and right, and step on the clutch pedal before the seedling conveying cam contacts the seedling conveying arm.

Step 4: Stop the engine and place the main shift lever in the "Neutral" position.

Step 5: Loosen the fixing bolts of the transverse seedling collection adjusting lever, and move the lever to a suitable position.

Step 6: Tighten the fixing bolts.

10) Adjustment method of transplanting depth.

As shown in Fig. 3-47, the depth of rice transplanting can be selected by using the transplanting depth adjusting lever (nine gears in total): ① is the deepest and ⑨ is the shallowest in Fig. 3-48; ⑦–⑨ are generally not used, because the seedlings may lodge or float if too shallow. Generally, the rice transplanting depth is 0.5–1 cm (Fig. 3-49).

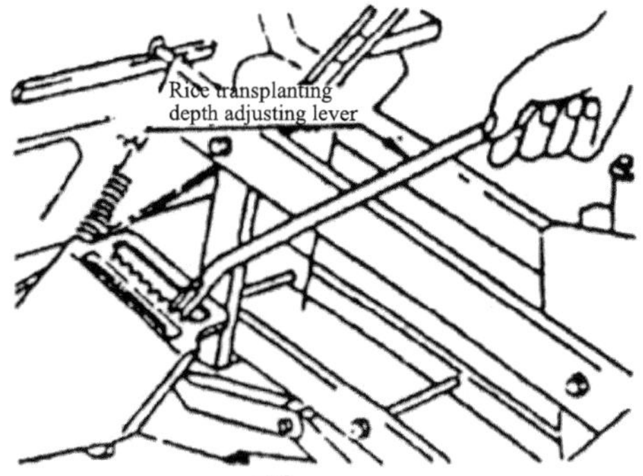

Fig. 3-47 Rice Transplanting Depth Adjusting Lever

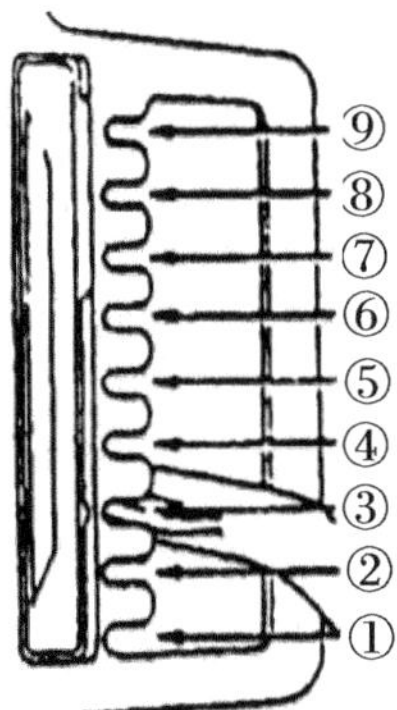

Fig. 3-48 Enlarged View of Rice Transplanting Depth Adjusting Lever

Fig. 3-49 Schematic Diagram of Rice Transplanting Depth

11) Adjustment method of longitudinal seedling collection amount.

Select the amount of seedlings to be collected longitudinally by adjusting the seedling collection amount adjusting lever left and right. The seedling collection amount increases when adjusting leftward, and vice versa. When the adjusting lever is placed in the standard position, the seedling collection amount is 11 mm (Fig. 3-50 and Fig. 3-51).

12) Adjustment method of longitudinal seedling conveying amount of seedling conveying belt.

The longitudinal seedling collection amount should be coordinated with the

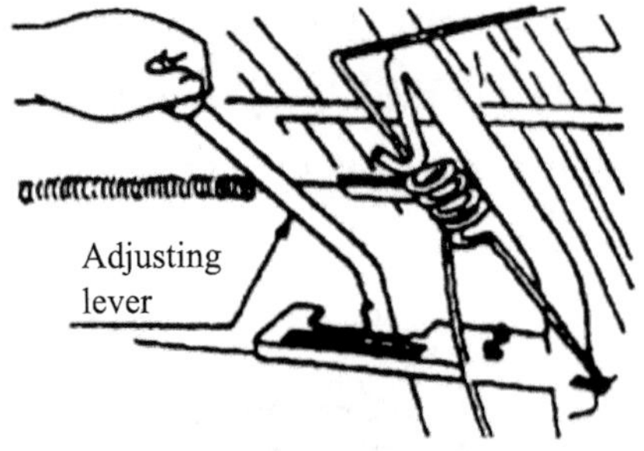

Fig. 3-50 Longitudinal Seedling Collection Amount Adjusting Lever

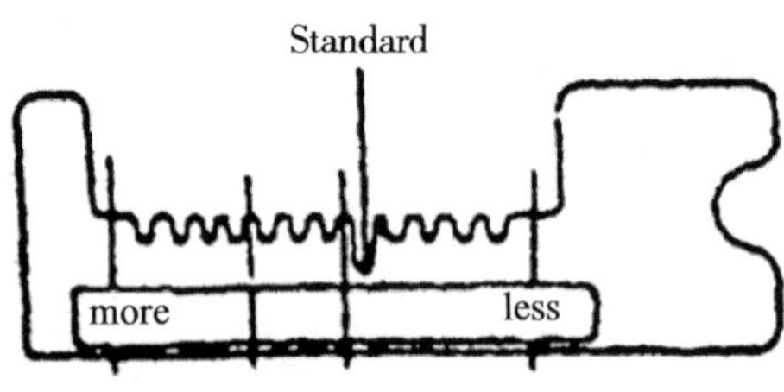

Fig. 3-51 Enlarged View of Longitudinal Seedling Collection Amount Adjusting Lever

longitudinal seedling conveying amount of the belt. There are 3 pin holes on the seedling conveying adjusting lever, and the positions of the pin holes inserted are suitable for different sections of the longitudinal seedling collection amount, as shown in Fig. 3-52. Pin holes ①, ②, and ③ are correspondingly suitable for ①, ②, and ③ (new models have linkage devices corresponding to the longitudinal seedling conveying amount and seedling collection amount, and no further adjustment is required).

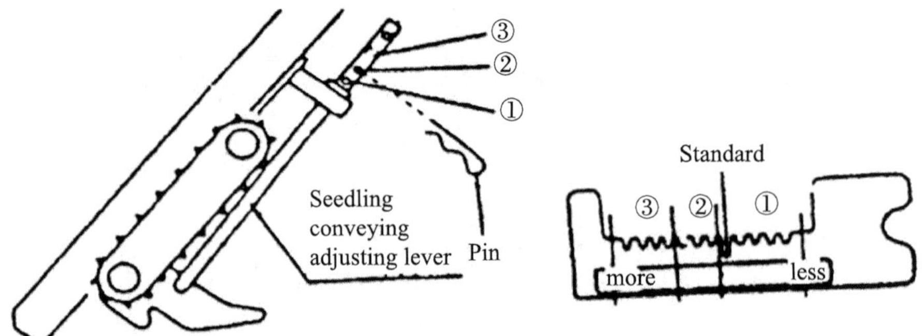

Fig. 3-52 Adjustment Method of Rice Seedling Conveying Belt

13) Adjustment method of hydraulic sensor.

According to the hardness of the soil surface of the rice transplanting field, an adjusting lever is used to change the sensitivity of the hydraulic sensor, which is placed at each position to realize soft-field or hard-field rice transplanting operations. When the adjusting lever is adjusted, the rice transplanting depth will change. After the adjusting lever changes, it is necessary to confirm the rice transplanting depth again. If the transplanting depth is below the required standard, the rice transplanting adjusting lever needs to be turned again to meet the requirements of shallow transplanting (Fig. 3-53).

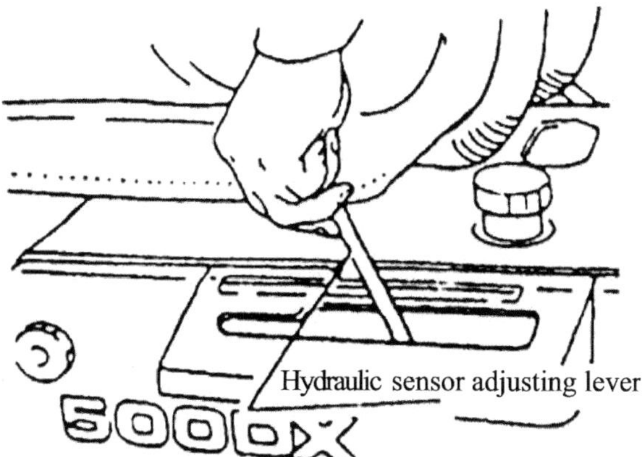

Fig. 3-53 Adjustment Method of Hydraulic Sensor

14) Adjustment method of blocking rod.

Changing the position of the blocking rod on the base plate (4 holes) can adjust the transplanting posture of the seedlings. Hole ③ is the standard position for installation. When the transplanting posture is not good, the position of the blocking rod can be changed according to Fig. 3-54, and the base plate can also be adjusted forward and backward according to the height of the seedlings. Note: During adjustment, the left and right sides should be consistent.

15) Operation steps of rice transplanting.

Step 1: After entering the paddy field, place the main shift lever in the "Neutral" position.

Step 2: Unhook the left and right brake pedals.

Step 3: Set the scribing indicator to the "working" position.

Step 4: Set the side indicator and the front plate guide to the "Working" position.

Step 5: Place the main shift lever at the "PTO" position and the rice transplanting clutch lever at the "Connected" position. Then, place the seedling box on the left and right ends of the machine, and place the rice transplanting clutch lever in the "OFF" position.

Step 6: Place the seedlings unloaded from the seedling collection board on the seedling box (Note: The seedlings can only be placed after the seedling box is placed at the left and right ends).

Step 7: Place the seedling preparation box on the auxiliary seedling rack. Note: A seedling collection board is placed on the auxiliary seedling plate.

Step 8: Place the main shift lever in the "1" position and the auxiliary shift lever in the "Low speed" position.

Step 9: Place the rice transplanting clutch in the "Connected" position.

Step 10: After the engine is set to "Medium speed", slowly release the clutch pedal and start rice transplanting. Before rice transplanting, check whether all devices have been adjusted. At the high field edge, place the front plate protection lever in the "Working" position, and at the low field edge, place it in the "Low field edge" position.

Specification:
After the clamp is removed,the position of the blocking rod can be changed.
There are four positions available.
There are also several positions available for floating plates.
At the time of leaving the factory,the position of the hole is ③.

Position A	Position B
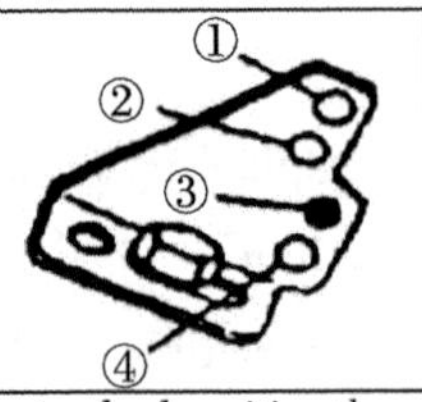 ① ② ③ ④	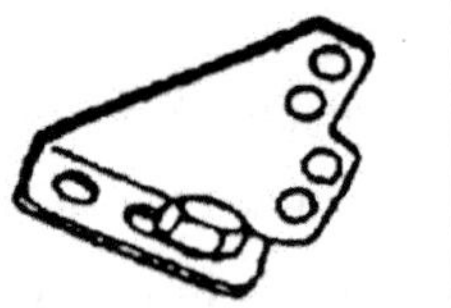

If the effect is not good after trial transplanting at the standard position,the adjustment can be made according to the following steps.

Problems	The position of the blocking rod should be changed to
The transplanted rice seedlings fall frward.	①②
The transplanted rice seedlings fall backward.	④
The transplanted rice seedlings are scattered.	④
The rice seedlings cannot be separated.	①②
The leaves of rice seedlings are scattered.	④

Fig. 3-54 Adjustment Method of Blocking Rod

16) Method of turning.

Step 1: When it is close to the field edge, make sure to place the auxiliary shift lever in the "Low speed" position, and the rice transplanting clutch lever in the "Up" position.

Step 2: Step on the brake pedal in the turning direction.

Step 3: Use the center lever and the side indicator to adjust the row spacing. The machine will drive in the same direction as the operation (Fig. 3-55).

Step 4: Place the rice transplanting clutch lever in the "down" position, remove the rice transplanting device and mark the line. Use the scriber on the other side when returning.

17) Use of automatic indicator (Fig. 3-56).

The scribing indicator is switched left and right as the rice transplanting device rises and falls. After the automatic indicator switch is turned on, the switch light and side indicator lamp will light up. Attention should be paid to placing the rice transplanting clutch lever in the "Up" position. After the rice transplanting device is completely lifted and the change of lamp of the scribing monitor is confirmed, the rice transplanting device is put down. After the rice transplanting device is fully lifted, if the automatic indicator switch is turned to "OFF", the side scriber will not work (Fig. 3-57).

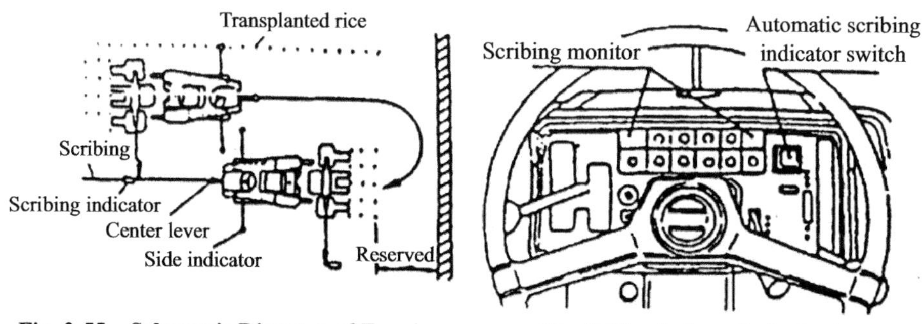

Fig. 3-55 Schematic Diagram of Turning Method

Fig. 3-56 Automatic Indicator

Fig. 3-57 Operation Method of Indicator Switch

18) Method of ending rice transplanting.

During the penultimate row of transplanting, use the group rice transplanting clutch lever and the seedling retaining ring when adjusting the number of rows according to the distance left from the ridge. Place the group rice transplanting clutch levers in the "OFF" position as required (Fig. 3-58). After the group rice transplanting clutch lever is used, it must be put back to the "Connected" position in the last row of transplanting. In stopping transplanting in a single row, the seedlings should be pulled up and fixed with retaining rings (Fig. 3-59). After using the seedling retaining ring, please ensure that the stuff returns to its original position.

When the safety clutch works and the rice transplanting arm stops working and makes a sound, the following measures should be taken: Disconnect the clutch immediately; disconnect the rice transplanting clutch lever and stop the engine; check whether there are stones, branches, and other sundries between the seedling collection outlet and the seedling needle, or between the rice transplanting arm and the floating plate; check whether the rice transplanting arm can rotate, whether it rubs against the front plate, and whether the seedling needle is deformed.

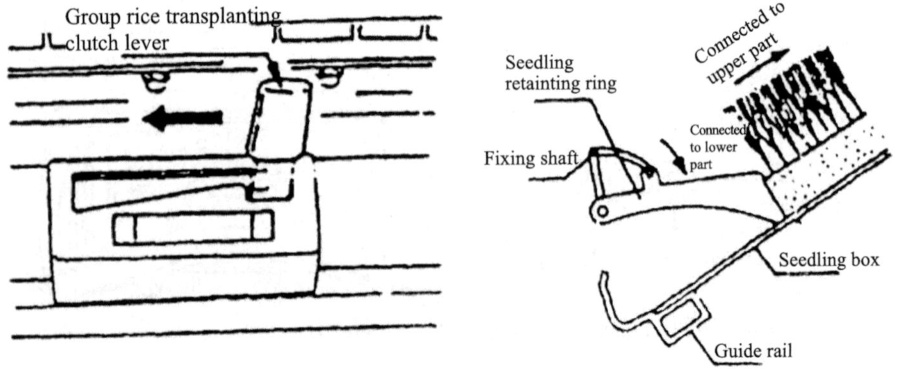

Fig. 3-58 Group Rice Transplanting Clutch Lever

Fig. 3-59 Adjustment Method of Seedling Retaining Ring

3. Potted seedling transplanter

(1) Model and features.

At present, the potted seedling machines used in Jiangsu Province are mainly

2ZB-6(RX-60AM) and 2ZB-6A(RXA-60T) potted seedling riding-type high-speed transplanters produced by Changzhou AMEC Machinery Equipment Co., Ltd. (Table 3-5).

Table 3-5　Main Performance Parameters of Potted Seedling Riding-type High-speed Transplanter

Model		2ZB-6(RX-60AM)
Overall dimensions	Overall length (mm)	3,140
	Overall width (mm)	18,901 (29,601 at work)
	Overall height (mm)	1,420 (2,010 at work)
Structure mass (kg)		537
Engine	Model	GH340
	Type	Vertical air-cooled 41-stroke gasoline engine
	Total displacement (cc{L})	340{0.34}
	Horsepower/Speed (kwcps/min)	5.9 (8.0) /3,600 [max. 8.3 (11.31) /4,000]
	Fuel tank (L)	9
	Starting mode	Electric start
Traveling wheel	Steering mode	Ackerman system (hydraulic)
	Structure	Hub rubber
	Diameter (mm)	670
	Structure	Hub rubber
	Diameter (mm)	900 (auxiliary wheel 700)
	Number of gears (gear)	(Rice transplanting 1) Backward 1
Rice transplanting part	Number of work rows (rows)	6
	Row spacing (mm)	330
	Distance between holes (mm)	120, 130, 140, 150, 160, 170, 180, 200, 220, 240

continued

Model		2ZB-6(RX-60AM)
Others	Rice transplanting depth (mm)	10–40
	Operating speed (m/s)	0–1.2
	Operating hourly productivity (hm^2/h)	0.26–0.37

The potted seedling riding-type high-speed transplanter (rice transplanter) is an original rotary slideway type sparsely planting technique that does not hurt seedlings. It enables five key working steps of "ejection-receiving-placing-feeding-planting" in the seedling planting process, and realizes the accurate cooperation of time intervals between transmission mechanisms; that is, placing the seedling tray longitudinal transfer claw from the seedling table and pulling it in, then the seedlings are ejected from the seedling tray–the seedlings are accurately received by the seedling receiver–the seedlings are placed on the conveyor belt–the rotating slideway mechanism completes the planting of potted seedlings, which breaks through the product structure principle of conventional rice transplanters. Especially from power output to various transmission parts, longitudinal transfer claw parts, seedling ejection parts, seedling receiver parts, seedling transfer parts, planting parts, and other moving parts, the technical characteristics of high accuracy and coordination ensure the advancement of product technology.

2ZB-6(RX-60AM) potted seedling single-ride 6-row transplanter is suitable for small field operations, and 2ZB-6A(RXA-60T) potted seedling double-ride 6-row transplanter is suitable for large field operations. The potted seedling transplanting technique can be carried out by shallow tillage and leveling (Fig. 3-60). The row spacing of the potted seedling riding-type high-speed transplanter is 33 cm, and the plant spacing is 12.4–28.2 cm, which can be freely adjusted to ensure density flexibility during transplanting. Planting depth: By changing the position of the planting depth adjusting lever, four different planting depths of 10–40 mm can be selected, and seedling alarm devices are installed to prevent missing planting. The potted seedling transplanter technique is most suitable for seedlings with a foliar age of 4.5–5.5 leaves and a seedling height of 12–30 cm (the period of seedling culture is 35–40 days in cold areas, 35–40 days in early warm areas, and 25–30 days in southwest warm areas). The foliar age of conventional carpet seedling culture is only 3–3.5 leaves, and the period of seedling culture is generally 15–20 days. Therefore, the flexibility of seedling age is low and the seedlings are weak. However, the transplanting technique of potted seedlings features few seedling age restrictions, high flexibility, and a large planting time span, which makes up for the deficiency of carpet seedling machine transplanting that cannot meet the requirements of double-cropping rice and multi-cropping rice due to the limitations of foliar age, seedling height, and transplanting time, and can ensure its high yield and yield growth. Comparing the above two, the complete set of potted seedling machinery and equipment has its own core advantages. It has wide popularization and application value in rice planting areas in northern China, rice-wheat dual-cropping areas in the middle and lower reaches of the Yangtze River, and hybrid rice application areas in southern China, promoting the increase in grain yield and ensuring food security.

Fig. 3-60 Potted Seedling Riding-type High-speed Transplanter 2ZB-6 (RX-60AM)

(2) Operating technique.

The operation technique is basically the same as that of carpet seedling transplanter, and will not be detailed here.

II. Preparation for Field Tillage

Tillage and leveling before rice field planting is an important part of the rice high-yield culture technique, generally including plowing, stubble removal, sunning, fertilization, soil crushing, harrowing, leveling, and other operations. Medium and small seedlings are used for machine-transplanted rice seedlings, which have relatively high requirements for the quality of field tillage and leveling and the application of basal fertilizer. The quality of tillage is not only directly related to the operation quality of the rice transplanter, but also to the early growth and rapid development of rice transplanted by the rice transplanter. Therefore, it is very important to carry out fine tillage and leveling of machine transplanting seedling fields.

1. Land preparation

(1) Quality requirements for tillage and leveling of machine-transplanting fields.

General requirements of machine rice transplanting for the field include: the rotary tillage depth is 10–15 cm, and the plowing depth is 12–15 cm, without repetition or leakage; the field is flat without stubble, the height difference is not more than 3 cm, the hardness of the topsoil is moderate, and the depth of the mud foot is less than 30 cm; slurry should not be settled until the slurry is clear, with a slurry depth of 5–8 cm and a water depth of 1–3 cm.

First, the field needs to be flat. Under the condition of a 3 cm water layer, seedlings should not be exposed and submerged, so that they can green up and grow neatly. Otherwise, the plantlets will dry up due to lack of water at high places, and the plantlets will be flooded due to the water depth in low-lying areas. Generally, it is required that the surface layer of the field is slightly muddy after tillage and leveling, the lower soil block is finely crushed, and the hardness of the surface soil is moderate (measured with a taper meter, and the depth of the cone top penetrating the soil layer is 8–10 cm). Although the high-performance rice transplanter has the advantages of multi-wheel drive and good paddy field passing performance, too-deep tillage layer will lead to the increased load of the rice transplanter, difficulty in walking and even slipping, which cannot ensure normal planting density. In addition, although the high-performance rice transplanter is equipped with a hydraulic profiling device to ensure that the machine has a low contact pressure, the land preparation times are too many, the soil layer is too sticky, which is not conducive to sedimentation, and there is still mud accumulation during the advance of the machine, which affects the planting quality.

Second, there must be no weeds, stubbles, or sundries in the field; otherwise, the stubbles and sundries will scrape down the transplanted seedlings during the machine's advancement.

(2) Tillage and leveling process.

1) Tillage and leveling of cropping land.

① Crushing of preceding crop straw: The preceding crop must be crushed and spread evenly when harvesting. If the preceding crop is wheat, the straw should be crushed during machine harvesting, and the stubble height should be less than 15 cm. If the straw is not crushed during machine harvesting, a straw crushing operation should be added or the straw needs to be removed from the field.

② Dry leveling: Under the condition of suitable soil humidity and moisture content, three methods can be used to eliminate stubble, i.e. forward (reverse) rotation, shallow tillage, and harrowing, among which reverse rotation is better. Avoid deep tillage as much as possible. During operation, the depth should be controlled within 15 cm, the plowing depth should be stable, the residue coverage rate should be high, and there should be no missing plowing. If the plot is uneven, a cross-dry leveling needs to be added to ensure that there are no blind ditches and potholes in the field, and that the height difference and flatness of the field meet the standard. For large-area field leveling, the laser leveling technique can be considered for dry leveling. If the laser land preparation technique is not available for the time being, the fields with large height differences have to be divided into grids and separated by small fields to achieve the quality standard of dry leveling within a relative range.

③ Water leveling: Shallow water should be poured in, and water leveling should be carried out after soaking for 24 hours. When the conditions are suitable, the soil can be dried to a moderate degree after dry leveling, and then soaked in water, so as not to form dead soil. Paddy field burial pulper, paddy field drive rake, and other equipment can be used for water leveling. In the process of water leveling, pay attention to controling the appropriate irrigation volume, not only to prevent the operation with rot, but also to prevent the operation with water shortage and rigidity.

Since the depth of rotary tillage and stubble removal before water leveling is shallower than that of the original tillage layer, and the operation conditions are complex due to slurrying and flattening, it is necessary to prevent different depths of mud feet and buried stubble from being brought out of the surface.

After water leveling, the surface of the field is flat and free of stubble, straw, and weeds. The depth of stubble is more than 4 cm, the depth of mud reaches 5–8 cm, and the height difference of the field is no more than 3 cm.

④ Sedimentation: After water leveling, the machine-transplanted field must be moderately compacted. Sandy soil must keep compacted for 1 day, sandy loam must keep compacted for 2–3 days, and clayey soil must keep compacted for 4 days before machine-transplanting. It is best to have the so-called "water spray" in the surface water layer of the field. Strictly prevent deep water and mud from damming during machine transplanting. For fields with high weed density, slurry sedimentation can be combined. After harrowing, suitable herbicides mixed with wet fine soil can be evenly spread, and a 6–10 cm water layer can be kept for 3–4 days for weed sealing and killing.

2) Winter plowing or winter plate stubble tillage and leveling.

After winter plowing and crop rotation, the field can be leveled by cross-cutting, harrowing, and scraping before water leveling. For untilled winter slack fields with few surface residues, shallow tillage or rotary tillage can be adopted for dry leveling before water leveling. For fields without stubble on the surface, with good winter tillage and leveling quality and flat ground, water leveling can also be carried out directly.

3) Tillage and leveling of winter paddy field.

In some areas, farmers cherish water, so they keep water in the field until early rice is planted after the preceding late rice is harvested. After being soaked in cold water for a long time, this kind of winter paddy field has poor soil permeability, deep mud foot, poor reducibility, and low soil temperature. For this kind of field, ditching needs to be made in advance in the field on sunny days to lift the field, and shallow rotary tillage needs to be carried out after fertilizer is applied. The field should be sunned and warmed on sunny days, and flattened and settled about 3 days before planting.

2. Basal fertilizer operation

Apply 1,000–1,500 kg of manure or 50–80 kg of 25% compound fertilizer per mu 5–10 days before transplanting to cultivate soil fertility; for fields with medium fertility, apply 30 kg of 35% rice special fertilizer or 40 kg of 25% compound fertilizer per mu as base fertilizer; fertilize first and then plow to achieve full-layer fertilization and soil-fertilizer blending.

III. Transplanting

1. Seedling age and specification

(1) Carpet seedlings.

The carpet seedling machine is used to transplant medium and small seedlings with soil. The seedling age of single-crop rice and middle-season rice is generally 15–20 days, and the seedling age of early rice is appropriately prolonged due to the low accumulated temperature. However, no matter how the seedling age changes, it is generally within 3.5–4.0 foliar age and transplanted when the seedling height is 12–17 cm. The quality of seedlings can be measured from two aspects: morphological and physiological indexes. In actual production, it can be judged by observing the morphological characteristics of seedlings. The main morphological characteristics of strong seedlings include thick and flat stem bases, emergent green leaves, multiple white roots, short and strong plants, and no diseased plants and pests. Among them, the thick and flat stem base is an important index to evaluate strong seedlings, commonly known as "sturdy seedling". For seedlings suitable for mechanized rice transplanting, in addition to strong individuals, there is an important overall index; that is, the quality of the seedling population is balanced. Conventional japonica rice seedlings are 1.5–3 seedlings per cm^2, and hybrid rice seedlings are 1–1.5

seedlings per cm^2. The root system of seedlings is developed, the number of white roots per plant is large, and the number of white roots per plant should exceed 10. The root system is firmly coiled. The thickness of the packing with soil is 2.0–2.5 cm, which is uniform in thickness and does not scatter after lifting. It is like a carpet, also known as a carpet seedling.

(2) Potted seedlings.

1.7–3.0 seedlings per hole for hybrid rice and 3–5 seedlings per hole for japonica rice are required for potted seedlings. The seedling age is about 30 days, the foliar age is 4.5–5.5, the seedling height is 15–20 cm, the stem base width of an individual is 0.3–0.4 cm, the average number of white roots per plant is 13–16 with 0.3–0.5 tillers, and the rooting power is 5–10 roots per 100 plants with a dry weight of more than 8.0g. The seedlings are uniform and neat, with stout stems, delicate and disease-free, and free of black roots and dead leaves. It has appropriate carbon and nitrogen nutrients, elastic leaves, a large number of root primordia, strong rooting power, and stress resistance after transplanting. It can take root early, survive early, and start seedlings early.

2. Technical standards for transplanting

(1) Transport and transplanting.

Machine-transplanted seedlings are transported and transplanted according to different seedling culture methods, so as to reduce the number of seedling movements, ensure the size of seedlings, prevent withering, and ensure that seedlings are transported and planted at the same time. In the case of high temperatures in the scorching sun, sunshade facilities should be provided during operation and release.

Soft (hard) tray seedling: If conditions permit, the seedlings can be transported to the field along with the tray, or the seedlings in the tray can be carefully rolled up after lifting and stacked on the seedling transport vehicle. Generally, the stacking is in 2–3 layers. Do not increase the pressure on the bottom layer due to excessive stacking to avoid deformation of the seedlings and breakage of the seedlings. After the seedlings are transported to the field, they are unloaded and placed horizontally immediately, so that the seedlings can be naturally stretched to facilitate machine transplanting.

(2) Plant spacing and row spacing.

1) Carpet seedlings.

The basic seedlings per mu of the field are determined by the row spacing, plant spacing, and the number of seedlings per hole. The row spacing of the rice transplanter is fixed at 30 cm, and the plant spacing is adjusted in multiple gears or steplessly. The corresponding planting density per mu is 10,000–20,000 holes. Correctly calculating and adjusting the number of planting and transplanting holes per mu and the number of plants per hole can ensure an appropriate number of crops may grow in the field. In actual production operations, it is generally necessary to determine the planting spacing beforehand, and then adjust the seedling collection amount of the seedling claws, or the number of plants per hole, in order to meet the requirements of agronomy for basic seedlings.

The rice transplanter adjusts the area of rice seedlings picked up by the claw according to the longitudinal seedling collection amount and transverse seedling feeding amount, thereby changing the number of plants per hole. For example, the TYM PF455S rice transplanter has a longitudinal seedling collection amount ranging between 8 mm and 17 mm, with a total of 10 gears. Each displacement by a magnitude represents a change of 1 mm. If the lever moves leftward, the seedling collection amount increases, and the amount decreases when the lever moves rightward. When the standard seedling collection amount (11 mm) is changed, it is necessary to use a seedling picking gauge for correction. The lateral movement adjustment device is located on the circular disk on the bracket of the planting part, and is marked at three positions, “26”, “24”, and “20”, indicating the movement of the seedling box by 10.8 mm, 11.7 mm, and 14 mm, respectively. After the horizontal and vertical matching adjustment, 30 different types of small seedling fields are formed, which are up to 2.38 cm^2 but not smaller than 0.86 cm^2. In general, the gear for lateral seedling picking can be fixed before the longitudinal seedling collection amount is changed by the lever. Under this principle, it is possible to adjust the seedling collection amount based on the density of seedlings to ensure a reasonable number of seedlings per hole.

In practical operations, the number of plants per hole should be inversely calculated based on the basic number of seedlings per mu and plant spacing in accordance with agronomic requirements. For example, there should be 60,000–80,000 basic seedling plants for a certain rice variety. If the plant spacing is 12 cm, it can be calculated that there are about 18,000 holes per mu, and about 3.5–4.5 plants per hole. In the meantime, attention should be paid to improving the uniformity of planting. Uniformity refers to the distribution of the actual number of seedlings planted and inserted in the hole. Because there is a certain automatic adjustment ability of tillering and ear formation, within the range of ± 1 for the planned number of seedlings, if the planned average number of seedlings in the hole is 3, it will be deemed uniform arranged when 2–4 seeding crops are actually planted and inserted per hole. Generally, the uniformity is required to be above 90%.

2) Potted seedling.

The row spacing of the potted seedling transplanter is 33 cm, and the plant spacing is 12.4–28.2 cm, which can be freely adjusted to ensure density flexibility during transplanting.

(3) Transplantation depth.

At the beginning of each operation, a certain length of transplantation should be done as an experiment, and the number of seedlings per hole and the depth of planting should be checked. This can not only timely adjust the seedling collection amount according to the density of seedlings to ensure there are 3–5 seedlings per hole, but also can timely adjust the planting depth according to the specific operating conditions of the field to make sure that the seedlings are not unsteady or tend to fall over, preferably exposed to the water surface. After the operating status meets the requirements and becomes stable, continuous operation can be started.

Section IV Water and Fertilizer Management of Machine-transplanting Paddy Fields

Water and fertilizer management measures for machine-transplanting paddy fields are generally consistent with those for manual transplanted rice seedlings, but they also have their own characteristics. Machine transplanting (Fig. 3-61) is suitable for small-and medium-sized seedlings that are younger and have poor stress resistance and a longer seedling recovery stage. However, the wide row and shallow transplanting of machine-transplanted rice facilitates the occurrence of low nodal region tillering. Their tillering happens explosively, and the tillering stage is also long. If the seedling stage is advanced, more peak seedlings will appear, resulting in a decrease in the spike rate and smaller spikes. In response to this growth characteristic of machine-transplanted rice, water and fertilizer management measures such as early stabilizing, in-process controlling, and late promoting can be adopted in production management to ensure earlier green-up, earlier tillering, and earlier field draining. In the middle and late stages, strict water and mud management can promote the formation of large spikes, in order to achieve high and stable yields and high-quality machine-transplanted rice.

Fig. 3-61 Machine Transplanting

I. Green-up and Tillering Stage

1. Growth characteristics during the green-up tillering stage

The green-up tillering stage refers to the period from transplantation to young panicle differentiation. This stage focuses on the growth of vegetative organs and is a key period for determining the number of spikes, as well as for laying the foundation for larger spikes, multiple spikes, and final high yield. In terms of production, reasonable cultivation measures should be used to shorten the green-up stage, promote early and sufficient tillering, stimulate more spikes, control ineffective tillering, cultivate strong tillers and large spikes, and accumulate sufficient dry matter.

Compared with conventional manual transplanting and seedling throwing, machine transplanting has its unique reproductive characteristics during this period, which are summarized as follows.

(1) Poor stress resistance.

The suitable transplanting age for machine-transplanted rice is 15–20 days. During these days, the foliar age is basically characterized by 3–4 leaves, and the seedling is 13–18 cm tall. The seedling is at or has just finished the weaning stage and enters the autotrophic stage. The number of root bases developed is relatively small. In addition, the planting density is high, and the root system is tightly coiled. During machine transplanting, the root system is cut and injured, and it takes days for the injured root system to recover and grow. After machine transplanting, the stress resistance of the seedling is weaker than that of conventional manual transplanting. Therefore, compared to conventional manual transplanting and seedling throwing, the green-up stage of machine-transplanted seedlings is longer, generally 3–5 days long, or even 5–7 days. Within 5 days after machine transplantation, seedlings will not substantially grow and will wait 14 days before tillering begins.

(2) Serious injury to seedlings.

During the conveying of rice seedlings, serious injury to seedlings is common, which, however, is often ignored by farmers. During the conveying of seedlings from the seedbed in the seedling field to the seedling box of the rice transplanter, the seedling blocks are rolled into tubes, stacked in multiple layers, and moved to the field manually or by using a handcart. Then, the seedling tubes are opened and placed onto the seedling box of the rice transplanter one by one. Due to the fragility of tender green seedlings, which are prone to breakage or rupture, the farther the transportation distance and the longer the stacking, the more serious the injury to the seedlings. After machine transplanting, it takes days for the seedlings to heal and grow, leading to a long green-up stage. In serious cases, seedlings may come to death lacking culms. Some people do not understand the essence of this phenomenon and blame the rice transplanter, which is wrong.

(3) Multiple tillering nodal regions and long tillering stage.

The foliar age of machine-transplanted seedlings is 1.5–2 positions younger than that of manual-transplanted seedlings. The effective tillering stage of manual-transplanted rice in the field has been extended by 2 leaf positions, about 7–10 days. Therefore, the number of tillering nodal regions has increased. In addition, seedlings are planted in wide rows and shallow depths, so they tend to benefit from superior plant temperature and light conditions, strong rooting ability, and low nodal region. As a result, manual-transplanted rice seedlings stand out for their explosive tillering and rapid increase. Therefore, it is easy to have too many peak seedlings, resulting in a drop of spike rate and a smaller spike type, which is worth much attention in cultivation management (Fig. 3-62 and Fig. 3-63).

Fig. 3-62 Severely Injured Seedlings

Fig. 3-63 Multiple Tillering Nodal Regions with Vigorous Tillering

2. Characteristics of water and fertilizer requirements at green-up and tillering stages

(1) Characteristics of water requirement.

The purpose of irrigation and drainage is to regulate the moisture content of paddy fields to fully meet the physiological and ecological water requirements of rice. In production practice, water and mud management should be based on the overall situation of the entire field, and flexible water management measures should be taken according to the weather and seedling conditions to speed up seedling survival and tillering.

During the green-up stage of rice, it is necessary to maintain a certain watercourse in the rice field to allow seedlings to grow in an environment with relatively stable temperature and humidity, and ensure they have new roots and green up as early as possible. Nevertheless, the watercourse must not go beyond the fully mature auricle on the top; otherwise, the recovery of growth will be hindered. Early planted seedlings should be irrigated with water shallowly during the day and deeply at night due to the low temperature. When cold waves come, it is advised to carry out deep irrigation to resist the cold and protect the seedlings. In the case of overcast rain during the green-up stage, shallow water or wet irrigation should be adopted. The suitable field moisture content for rice tillering is that the soil moisture is between highly saturated and shallow water to promote early tillering and rapid development. Tillering will be inhibited if the watercourse is too deep. Field drainage and sunning are often used to suppress ineffective tillers.

They are specific to the characteristics of short seedling age, small individual size, and weak growth of machine-transplanted rice. The principle of thin water irrigation and shallow water seedling survival should be adhered to. The dry environment facilitates rooting, while the moist environment facilitates seedlings. The time for the seedlings to green up can be shortened, provided that the root system of the seedlings develops well. During irrigation, it is necessary to prevent enduring deep water from causing a lack of oxygen in the root and bud of seedlings, resulting in a long green-up stage. In production, due to the small seedling size and small root system during machine transplanting of rice, farmers often focus more on seedling protection but neglect root promotion after transplanting, mostly adopting the method of water layer seedling protection, resulting in poor root growth and delayed tillering. This must be specially noted and avoided.

(2) Characteristics of fertilizer requirement.

Machine-transplanted rice selects a high-yield cultivation path of small population, strong individuals, and high accumulation. Due to the use of small- and medium-sized seedlings for transplanting, the ability to absorb fertilizer at the initial stage after transplanting is not as good as that of manual transplanting. Therefore, tillering fertilizer must be applied multiple times but in small quantities each time. At the early stage of machine-transplanted small seedlings, the root amount is small and the absorption capacity is weak. At the same time, the field tillering starts late and the tillering stage is long. Therefore, the basal fertilizer for machine-transplanted rice should be appropriately reduced, and the amount of tillering fertilizer should be increased accordingly. According to the test results in recent years, it is suggested that basal fertilizer should keep a proportion of 30% to 40% in base tiller fertilizer, or 60% to 70% in tillering fertilizer. At the same time, the application of tillering fertilizer will produce a better effect when the seedlings begin to have more roots. In other words, when the second inner leaf grows after planting, tillering fertilizer should be applied at different times, which can synchronize the fertilizer efficiency with the optimal tillering occurrence period, promote effective tillering, ensure the formation of spikes in an appropriate number, while controlling ineffective tillers. By doing so, large spikes will be produced, and fertilizer utilization efficiency will be improved.

If fertilization is done too early, i.e., applying tillering fertilizer immediately after planting. At this time, it is in the post-planting tillering stagnation period, with weak roots that cannot exert the efficiency of fertilizers. Fertilization immediately after planting inhibits root development, delays the tillering occurrence period, and causes insufficient spikes. On the contrary, if tillering fertilizer is applied too late, it will be likely to lead to a large population and a low spike rate at a high tillering stage. The number of grains per spike is small, though there are many spikes, and it is unlikely to realize a high yield. Hence, in the medium term, it is necessary to keep track of the seedling situation, timely drain water, sun the field, and control fertilization to inhibit ineffective tillering, thus avoiding high seedling numbers but low spike rate due to excessive fertilizer application.

3. Water and fertilizer operation from green-up to tillering stage

(1) Water and slurry management.

1) Transplanting to green-up stage.

In general, transplanting in shallow water and intermittent dehydration should be adhered to promote roots as appropriate after planting, and frequent and shallow irrigation should be adopted to promote tillering after complete seedling establishment. The specific operation is as follows: After the completion of machine transplantation, timely irrigation and seedling protection should be carried out, and the depth of the watercourse should also vary depending on the age of the transplanting seedlings. If the age is short, the watercourse should be shallow. If the seedling age is longer, the depth should be greater depending on the specific seedling height. The watercourse should be kept at about 1/2 of the seedling height. In the case of sunny weather, deep water should be irrigated to protect the seedlings, and the watercourse should be maintained at about 2/3 of the seedling height. On rainy days, a thin layer of water will be sufficient (Fig. 3-64). Three to four days after transplanting, seedlings should be managed in a thin watercourse. It is prohibited to keep them in deep water for a long time; otherwise, blackening of the root system and seedling heart will occur due to lack of oxygen, seedlings will grow slowly and leaves become yellowish, causing stiff seedlings or even rotten seedlings. After 3–4 days, the water should be promptly drained to expose the field for about 2 days to allow the soil to get ventilated and warmed and promote the production of new roots and seedlings. If land preparation is not appropriately finished, the field is likely to have a large drop. Seedlings can hardly root in deep water at low places, while seedlings at high places may even be killed due to drought.

Fig. 3-64 Seedlings in a Thin Layer of Water (Green-up Stage)

2) Effective tillering stage.

Machine-transplanted seedlings have a small seedling size and need a relatively long survival time. Once the seedlings enter the tillering stage after survival, water and mud management should adopt frequent and shallow irrigation to keep the watercourse of the field fresh, which is conducive to placing tillering nodes on a more suitable soil surface layer. After natural drying, water can be added to adjust fertilizer with water, foster roots with air, and coordinate water and air, which is conducive to low-level tillering and achieving early tillering with fresh water. During the entire effective tillering stage, it is advisable to maintain a shallow watercourse or adopt wet irrigation methods.

The specific frequent and shallow irrigation method is as follows (Fig. 3-65). Paddy fields of clay soil texture or medium-low fertility are irrigated with 1–3 cm shallow water once, kept for 3–5 days, and then dried naturally. When there is no surface water in the field and the soil is wet, the field is irrigated with shallow water once again. The process is repeated, so as to achieve a growth environment in which the aboveground part is coordinated with the below ground part, the root system is coordinated with the growth of leaves and tillers, and the population is coordinated with the individual. For fields with clay soil, high fertility level or high fertilizer application, the wet irrigation water management method of daytime watering on sunny days and nighttime field exposure is adopted, which can effectively promote the early growth and rapid development of tillers and help promote the formation of full spikes and large spikes.

Fig. 3-65 Frequent and Shallow Irrigation (Tillering Stage)

3) Ineffective tillering stage.

The number of basic seedlings planted in a unit area of machine-transplanted rice is slightly more than the number of seedling holes in conventional cultivation. From the perspective of population development, survival and tillering take slightly longer. Once the tillering starts, it shows a strong potential for seedling emergence; stems and tillers of the population increase rapidly, and the peak seedlings come vigorously. If the number of peak seedlings in the population is not controlled properly, it is easy to overgrow. Therefore, the field should be drained early at the right time to control ineffective tillering and improve the quality of the population.

The suitable field draining period mainly depends on the number of stems and tillers in the field. Field draining at enough number of seedlings means draining the field when the total number of tillers in the field reaches the expected number of spikes (commonly known as sufficient seedlings). Also, the principle of "timing prevails over seedlings" is applied, that is, field draining is carried out when the critical foliar age for effective tillering is reached, regardless of the total number of stems and tillers in the field. Compared with conventional cultivation, the seedling stage of getting enough tillers and the peak seedling stage can be advanced by about one foliar age. Generally speaking, the field draining is started by natural water cut-off and drying when the number of population tillers reaches about 80% (70%–90%) of the expected number of spikes, following the principle of "early draining, gentle draining, and draining for multiple times".

Excessive field draining should be avoided, and gentle draining should be done in stages. After 3–4 repetitions, the field should be drained until there are narrow slits on the field edge, feet don't get trapped in the mud, the soil does not turn white, the leaves are hard and straight, and the leaves fade to yellow. To suppress the occurrence of ineffective tillering, the elongation of the basal internodes can be controlled to keep the peak seedlings at 1.3–1.4 times the effective number of spikes, so as to improve root system vigor, enhance stalk elasticity, and improve resistance to lodging and diseases (Fig. 3-66 and Fig. 3-67).

Fig. 3-66 Before Field Draining (1 Foliar Age before Effective Foliar Age)

Fig. 3-67 After Field Draining (Before Jointing)

(2) Scientific fertilization.

Generally, a green-up and tillering fertilizer is applied to small seedlings within 5–7 days after planting. A urea of 5–7 kg/mu is applied after 15:00. There should be a water layer in the field to avoid seedlings burning from ammonia. After fertilization, water should be conserved for 5–7 days, and at the same time, the leveling gap should be made to prevent the rainwater from flooding the bud of the seedlings. About 10–12 days after planting, another 7–9 kg of urea should be applied per mu to meet the needs of sufficient tillering of machine-transplanted rice. About 18 days after planting, a balanced fertilizer should be applied depending on the seedling condition. Generally, 3–4 kg of urea or 9–12 kg of 45% nitrogen, phosphorus, and potassium compound fertilizer is applied per mu. The fertilizer is applied at the proper amount to make sure that the leaf can fade out in time before the effective tillering foliar ages and the leaf color fades normally. For the field with sufficient fertilizer and early and fast seedling development, fertilizer application should be postponed a bit and compound fertilizer should be applied mainly. For fields with poor ridges and slow-growing seedlings, fertilizer should be applied early and heavily to promote the early generation and fast development of tillers.

(3) Preventing stunted seedlings.

Stunted seedlings represent a kind of abnormal growth condition during the tillering stage of machine-transplanted rice, mainly manifested as slow growth of tillers, rice bushes clusters, stiff leaves, growth stagnation, root system growth obstruction, and other phenomena. The causes of stunted seedlings are complex and of many types, mainly caused by fertilizers, water, and pesticides. Stunted seedlings caused by fertilizer and water are mainly manifested in repeated top dressing and seedling burning by ammonia or long-term flooding; stunted seedlings are caused by improper herbicide species and dosage in the weeding process and water flooding of inner leaves after the pesticide. Effective prevention and control should be carried out for different causes.

II. Jointing and Pregnancy Stage

1. Growth characteristics in the jointing and pregnancy stage

During the jointing and pregnancy stage, the growth of vegetative organs, such as the last few leaves and the root system, is completed with the stem as the center, while reproductive growth centered on the young panicle differentiation takes place. This stage is important not only for keeping tillers and increasing spikes, but also for protecting and increasing flowers and increasing grains, as well as laying the foundation for grain filling. Due to the accelerated growth of the organs, the accumulation of plant dry matter accounts for about 50% of the lifetime dry matter accumulation in just 30 days. During the jointing and pregnancy stage, rice needs more water and fertilizer and it is most sensitive to external environmental conditions in this stage. Therefore, the cultivation goal of this stage is to strengthen the stems and roots and promote large spikes and sufficient grains on the basis of strong seedlings and tillers in the early stage to prevent excessive growth and lodging, and to create good conditions for later grain filling.

2. Characteristics of water and fertilizer requirements in the jointing and pregnancy stage

(1) Characteristics of water requirement.

The jointing and pregnancy stage is the period when rice needs the most water in its life. In particular, the meiosis stage in pollen mother cells is particularly sensitive to moisture content and is the critical period of water requirement. In addition, the amount of seepage from the paddy field has increased after the field drying and re-watering, and the water requirement in this stage generally accounts for 30%–40% of the whole growth period.

(2) Characteristics of fertilizer requirement.

The fertilizer applied during the pregnancy stage of rice is called spike fertilizer, which is good for both preserving the number of spikes and striving for large spikes. However, excessive growth of leaf area should be prevented in order to form a well-configured canopy structure. This can not only expand the "library" to form a higher total number of glumous flowers, but also strengthen the "source" and smooth the "flow" to form a higher grain-leaf area ratio, so as to help improve the percentage of filled grains and thousand-grain weight. According to the application time and effect, spike fertilizer can be classified into flower-promoting fertilizer and flower-preserving fertilizer.

Flower-promoting fertilizer is a fertilizer that encourages the differentiation of branches and glumous flowers. Flower-preserving fertilizer is a fertilizer that prevents the degradation of glumous flowers and increases the number of grains per spike. It is also very effective in preventing premature senescence of rice in the late stage, improving the percentage of filled grains, and increasing the grain weight. After effective control of peak seedlings, good spike fertilizer should be applied in due time according to the seedlings to facilitate the formation of strong stems and large spikes, which can optimize the medium-term growth and optimize the population structure.

3. Water and fertilizer management in the jointing and pregnancy stage

(1) Reasonable irrigation.

Intermittent irrigation can be implemented after field drying and seedling control to keep the field surface moist. A proper water layer should be maintained to protect the seedlings during the pregnancy stage, the flooding depth should not exceed 10 cm, and the deep water layer should not be maintained for too long.

(2) Spike fertilizer application.

In general, the flower-promoting fertilizer is applied at the beginning of spike differentiation, i.e. when the remaining foliar age is 3.5 to 3.1. The specific application time and dosage depend on the seedling condition. If the leaf color fades normally, urea can be applied for 10 to 12 kg per mu. If the leaf color is darker and does not fade, the amount of fertilizer application can be delayed and reduced. If the leaf color is lighter, the flower-promoting fertilizer can be applied 3 to 5 days earlier, and the dosage may be increased appropriately. If the leaf color is darker, fertilizer may not be applied. Flower-preserving fertilizer is generally applied 18 to 20 days before head emergence, i.e., when the number of remaining foliar age is 1.5 to 1. The specific application period is determined by stripping to check the remaining foliar age of more than 10 simple stems. The flower-preserving fertilizer is applied when the remaining foliar age number of 50% of the effective stem tillers does not exceed 1.2. Generally, 4 to 6 kg of urea should be applied per mu, and if the leaf color is light and the population growth is small, more urea needs to be applied but the dosage should not exceed 10 kg per mu; if the opposite is the case, it should be applied less or not be applied (Fig. 3-68 and Fig. 3-69).

III. Grain Filling Stage

1. Growth characteristics of grain-filling stage

In this stage, the reproductive growth of the rice plant is dominant and the center of growth shifts from spike differentiation to the development of the grains. Physiological metabolism is dominated by carbon metabolism, and photosynthetic

Fig. 3-68 Appearance of Machine-transplanted Rice Seedlings in the Jointing and Pregnancy Stages

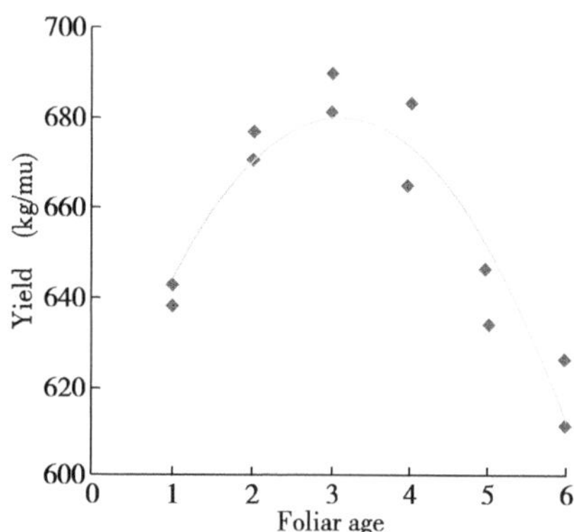

Fig. 3-69 Relationship between Foliar Age for Spike Fertilizer Dressing and Yield

products such as sugars made by leaves and nutrients stored in stems and leaf sheaths before heading are transported to the grains, which is a critical period to determine the percentage of filled grains and grain weight. The cultivation objectives in this stage are to nourish the roots and protect the leaves, prevent premature senescence, enhance the photosynthetic capacity of the rice plant, and improve the percentage of filled grains and grain weight, with the main purpose of preserving spikes, striving for grains, and increasing weight. High-yielding rice requires 3 to 4 green leaves to remain per effective tiller within 20 days after flowering (Fig. 3-70).

Fig. 3-70 Appearance of Machine-transplanted Rice Seedlings in the Grain-filling Stage

2. Characteristics of water and fertilizer requirements in the grain-filling stage

(1) Characteristics of water requirement.

During the heading and flowering stage, the sensitivity of the plant to water deficiency in the paddy field is second only to that of the booting stage (Fig. 3-71). In the case of drought, the serious situation involves difficulty in heading and flowering, while the less serious situation involves the impact on pollen and stigma vigor and an increase in the empty husk rate. When flowering and fertilization are completed, rice enters the grain-filling stage, a period when it is necessary to ensure moisture content supply but not long-term flooding. If the soil is lacking in water, it is not conducive to grain-filling, and it also has adverse effects on the speed of filling and rice quality; if it is flooded for a long time, the root system is not vibrant, the leaves show premature senescence, and the blighted grains increase. Therefore, the moisture content management method of intermittent irrigation is mainly used to keep the soil moist. It satisfies physiological water demand, but also maintains soil compaction without softening, enhances soil root aeration, maintains root system health, delays vitality decline, and prevents bacterial wilt and premature senescence, so as to achieve the purpose of regulating the air with water, nourishing the roots and preserving the leaves, and strengthening the grains by wet-and-dry irrigation.

Fig. 3-71 Heading and Flowering of Machine-transplanted Rice (Grain-filling Stage)

(2) Characteristics of fertilizer requirement.

The fertilizer applied around the time of rice heading is called granular fertilizer. The granular fertilizer can increase the nitrogen concentration in the upper leaves, improve the protein content of the grains, and delay leaf senescence and improve root system vigor, thus increasing the filling material and increasing grain weight.

After full heading, the dressing of granular fertilizer should be well-timed. It should not be applied if the seedlings do not turn yellow, or there is much rain and little sunshine, or there is a disease. The amount of fertilizer applied should not be too much, and top dressing can be used. After heading, it is also necessary to control sheath blight, blast of rice, rice leafroller, rice planthopper, etc.

3. Water and fertilizer management in the grain-filling stage

(1) Water and slurry management.

From the head emergence to the following 20 to 25 days, the rice plant needs more water. The shallow water layer should mainly be maintained; that is, after irrigation with shallow water, the shallow water layer should be filled again when there is water if the field is stepped by foot after natural depletion. During the milky stage, intermittent irrigation should be applied to keep the field moist, but water should not be cut off. During the dough stage, irrigation can be carried out in the way of "passing water", so that the paddy field is in a state of alternating between waterlogging and water dropping. The drying period is gradually lengthened, and the amount of irrigation is gradually reduced until the water is cut off one week before harvest, so as to ensure that the grains are fully filled, which can not only improve the yield but also improve the quality.

In production, it is important to prevent early water cut-off and early cutting of green crops before maturity.

(2) Granular fertilizer application.

Granular fertilizer is "applied according to the seedlings. After the head emergence of rice, fertilizer is generally not needed. If the leaf color is obviously yellow, 1% urea can be used with 0.2% potassium dihydrogen phosphate solution for foliar feeding 1–2 times. This not only replenishes the fertilizer but also avoids reluctant ripening and late maturity, thus increasing grain weight.

IV. Precise Irrigation and Fertilization Techniques in the Whole Growth Period

1. Precise irrigation techniques

Previous studies have shown that the order of sensitivity to soil moisture content in various growth periods is peak tillering stage > germ cell formation stage> branch differentiation stage > late tillering stage > pollen formation stage, early grain-filling stage > late grain-filling stage. That is, the most sensitive period to water stress during the lifetime of machine-transplanted rice is around the peak tillering stage and meiosis stage. Except for these two sensitive stages, the highest yield is achieved in the late tillering stage, pregnancy stage, heading stage, and grain-filling stage with soil water potential of -15 kPa. It shows that high-yielding machine-transplanted rice is not meant to be irrigated with water layers for long periods of time, and alternating wet-and-dry irrigation throughout the whole growth period is more conducive to better yield.

Based on the experiment and large-area practice in various regions, the precise irrigation pattern of high-yielding machine-transplanted rice in its whole growth period is shown in Fig. 3-72.

Foliar age / Water layer	4	5	6	7	8	9	10	(11)	12	13	△14	15	16	17	Booting	Heading	Grain-filling stage	Harvesting
12, 9, 6, 3, 0, -3, -6, -9, -12, -15	Thin-water trans-planting	Exposing the field for moisturizing		Establishment of shallow water layer				Timed early field draining			Maintaining intermittent wet irrigation and alternate wet and dry condition							Water cut-off

Fig. 3-72 Precise Irrigation Pattern of High-yielding Machine-transplanted Rice

The specific measure is thin-water transplanting. After transplanting, appropriate intermittent dehydration is required to promote rooting, and after complete seedling establishment, frequent and shallow irrigation should be adopted to promote tillering. When the total number of tillers in the group reaches 80% of the expected number of spikes, the water supply should be cut off for natural draining. Gentle draining multiple times can effectively control the occurrence of ineffective tillers and improve the spike rate of tillers. One-time draining to a heavy extent should be avoided. After the jointing stage, shallow-wet alternate irrigation should be adopted; that is, to irrigate the field until the water level reaches 3 cm, then allow it to dry naturally until there is no water left in the high-yield ditch. After that, it will be re-irrigated. This process is to be repeated until one week before maturity. Such irrigation can not only meet the ecological and physiological water requirement of machine-transplanted rice, but can also facilitate the growth of strong stems and large spikes in the middle stage and the preservation of roots and leaves in the late stage, and significantly save water.

2. Precise fertilization techniques

The heavy use of fertilizer ensures a steady increase in rice yield and national food security. However, in the meantime, it brings negative effects such as low resource utilization efficiency, water and atmospheric environmental pollution, and deterioration of rice quality. The coordination between high yield and high efficiency and between high yield and high quality are the main scientific topics in the current theoretical research of rice cropping. Precise calculation of fertilizer consumption, fertilizer saving, and reasonable planning for fertilizer use are the most critical culture techniques in achieving the comprehensive goal of "high yield, high quality, high efficiency, eco-friendliness, and safety" in rice production.

The precise quantitative fertilization technique of rice, also known as soil testing and formulated fertilization technique, is a technique used to determine the appropriate dosages and ratios of nitrogen, phosphorus, potassium and micro-fertilizers and the corresponding fertilization techniques to be applied with organic fertilizers used, by calculating the appropriate dosage of nitrogen fertilizer required using the Stanford Equation, through comprehensive application of modern agricultural scientific and technological achievements, and based on the soil nutrient supply, fertilization requirements of crops, and fertilizer effect. The core is to resolve the conflict between the fertilizer requirement from crops and nutrient supply from the soil, and at the same time to supplement the required nutrient elements accordingly; that is, to supplement what is lacking and the necessary amount. This way, it would be possible to achieve the balance of all nutrients, satisfy the requirements of crops, improve the utilization rate of fertilizer, reduce fertilizer dosage, raise crop yield, enhance the quality of agricultural products, save labor, lessen expenditure, and increase income.

(1) Reasonable application ratio of nitrogen, phosphorus, and potassium fertilizers.

The maximum fertilizer efficiency and crop yield can only be reached with a balanced and coordinated uptake of nitrogen, phosphorus and potassium. The absorption ratio of nitrogen (N), phosphorus (P_2O_5) , and potassium (K_2O) in high-yielding rice is 1 : 0.45 : 1.2, which is a physiological index reflecting the balanced nutritional composition of the three elements.

Nitrogen use planning: Base tiller fertilizer to spike fertilizer ratio 6 : 4 for machine-transplanted seedlings is beneficial for increasing yields. Excessive nitrogen application in the early or later stages cannot achieve high yields. The reason is that when the ratio of base tiller fertilizer to spike fertilizer is 6 : 4, the population development is relatively coordinated, and the number of spikes and grains increases synergistically. Both the yield and the current nitrogen utilization efficiency (over 40%) are the highest. Therefore, this ratio is the optimal mode of nitrogen use planning. The proportion of basal fertilizer in base tiller fertilizer should be 30%–40%, and that of tillering fertilizer should be 60%–70%. Spike fertilizer should be applied twice, with 70% applied as the flower-promoting fertilizer during the inverse-3.5-leaf stage, and 30% applied as the flower-preserving fertilizer during the inverse-2-leaf stage.

Phosphate fertilizer should be used as basal fertilizer and applied all at once. Potassium fertilizer should be applied twice, with 50% first applied as basal fertilizer and 50% afterward as jointing fertilizer.

(2) Precise quantification of nitrogen fertilizer.

The precise quantification of nitrogen fertilizer should solve three issues: determining the total amount of nitrogen fertilizer to be applied, determining the proportion of base tiller fertilizer and spike fertilizer, and adjusting the application of spike fertilizer based on seedling conditions. The Solutions to these three issues are as follows.

1) Determination of the total amount of nitrogen fertilizer.

Using the Stanford Difference Method formula, the total amount of nitrogen fertilizer to be applied should be:

$$\text{General formula: Nitrogen application amount (kg/mu)} = \frac{\text{Target yield nitrogen requirement–Soil nitrogen supply}}{\text{Seasonal utilization rate of fertilizer nitrogen}}$$

① The nitrogen requirement for the target yield can be obtained by calculating the nitrogen requirement per 100 kg yield of high-yielding rice. The amount of nitrogen required per 100 kg yield in high-yielding fields varies in different regions, so the actual nitrogen uptake of local high-yielding fields should be measured.

$$\text{Formula ①: Nitrogen requirement for target yield (kg/mu)} = \frac{\text{Target yield} \times \text{Nitrogen uptake of 100 kg grain}}{100}$$

② The amount of soil nitrogen supply can be obtained by calculating the blank rice yield (the base yield) without nitrogen fertilization and the nitrogen requirement per 100 kg of rice yield. The formula fertilization by soil testing in different regions can provide a reference for determining the local soil nitrogen supply.

$$\text{Formula ②: Soil nitrogen supply (kg/mu)} = \frac{\text{Basic soil fertility yield} \times \text{Nitrogen uptake of 100 kg grain in nitrogen–free blank area}}{100}$$

③ The utilization rate of nitrogen fertilizer in the current season should be determined based on the utilization rate of nitrogen fertilizer under normal cultivation conditions (reasonable management of nitrogen fertilizer).

Note: The three main parameters for machine-transplanted rice and hand-planted rice are similar. For high-yielding japonica rice (about 700 kg), about 2.1 kg of nitrogen is required per 100 kg of rice, the seasonal utilization rate of nitrogen should be 40% or higher, and the nitrogen supply of soil can be determined based on the results of local nitrogen-free experiments.

There are many factors that affect the seasonal utilization rate of nitrogen fertilizer. However, our research has made it clear that in the same place, as long as attention is paid to avoiding excessive nitrogen fertilizer, preventing loss in fertilization methods, reasonably adjusting the proportion of base tiller fertilizer and spike fertilizer, and implementing reasonable “postponing nitrogen fertilizer”, it can be achieved with confidence that the utilization rate of nitrogen fertilizer in the current season will be increased to 40%–45% or even higher with less fertilizer, high efficiency and high yield.

2) Yield increase principle of “postponing nitrogen fertilizer”.

In postponing nitrogen fertilizer, the application ratio of base tiller fertilizer and spike fertilizer is adjusted from 10 : 0–8 : 2 to 5.5 : 4.5 (6 : 4–5 : 5) and 6.5 : 3.5 (7 : 3–6 : 4), which is an extremely important quantitative index for precise nitrogen fertilizer application. Significant fertilization reform occurred when traditional slow-acting farmyard manure shifted to fast-acting chemical fertilizers. The yield increase principle is briefly described as follows:

① Base tiller fertilizer. It mainly provides nutrients for the occurrence of effective tillering. When there are sufficient seedlings at critical foliar age for effective tillering, the nitrogen supply for soil should be weakened to promote the leaf color of the population becoming “yellow”, effectively control ineffective tillering and leaf elongation, delay row sealing, improve the light receiving conditions of the middle and lower leaves of the population from jointing to heading stage, improve the spike rate, and realize balanced development of underground and aboveground parts, to create perfect conditions for forming large spikes in the pregnancy stage.

If the proportion of nitrogen fertilizer in base tiller fertilizer is too large, the leaf color cannot become “yellow” normally at the ineffective tillering stage, which will result in vigorous growth in the middle stage; the row sealing will also be greatly advanced, the middle and lower leaves will be seriously shaded, and the high-yield population will be destroyed, which will bring a series of adverse consequences such as a sudden drop in spike rate, dysplasia of roots and stems, and serious diseases. The nitrogen absorption and utilization rate of base tiller fertilizer is low, generally only about 20%. The more such fertilizer is applied, the lower the utilization rate will be. Appropriate reduction of the application proportion of base tiller fertilizer can improve the utilization rate of nitrogen fertilizer in the current season.

Note 1: For the variety with more than 5 main stem elongation internodes (n) and a total foliar age (N) of more than 14 leaves, its critical foliar age for effective tillering is as follows: When medium and small seedlings are transplanted, the critical foliar age is N–n; when large seedlings with a foliar age of 8 and above are transplanted, the critical foliar age is N–n+1. Taking the variety with a total foliar age of 17 leaves and 6 main stem elongation internodes as an example, when medium and small seedlings are transplanted, the critical foliar age for effective tillering is N–n = 17–6 = 11, i.e. the foliar age is 11; when large seedlings are transplanted, the critical foliar age for effective tillering is N–n + 1=17–6 + 1 = 12, i.e. the foliar age is 12.

Note 2: From the ineffective tillering stage to the jointing stage, the leaf color of the population should become "yellow", and the top 4th leaf should be lighter than the top 3rd leaf (top 4th < top 3rd). The top 3rd leaf refers to the 3rd leaf counting downward from the top extending leaf, and the top 4th leaf refers to the 4th leaf counting downward.

② Effect of spike fertilizer. Applying spike fertilizer on the basis of yellowing in the middle stage can not only significantly increase the formation rate of large spikes, but also improve the wavering-tillering spike rate to ensure full spikes. The unit production efficiency of spike fertilizer is the highest, which is the most efficient fertilization in rice life. Appropriately increasing the application proportion of spike fertilizer is the key measure to increase yield.

③ There must be a reasonable proportion for postponing nitrogen fertilizer. For the variety with 5 elongation internodes, its nitrogen uptake before jointing only accounts for about 30% of the lifetime, and about 50% at the pregnancy stage, so the proportion of spike fertilizer can be increased to 40%–50%.

For the variety with 4 elongation internodes, its nitrogen uptake before jointing already accounts for 50% of the lifetime, so the proportion of spike fertilizer can only be increased to 30%–40%.

④ Postponing nitrogen fertilizer is a universally significant technique for increasing yield. Under the same level of nitrogen application, the results of the comparison test between the technique of postponing nitrogen fertilizer and local practice fertilization mode always can show a significant yield increase effect with a stable number of spikes, high spike rate, and obviously enlarged spike type. In 2006, 34 pairs of comparison tests between proportions of 6 : 4 and 8 : 2 (nitrogen application level: 12–14 kg/mu) were set up in 4 areas of Guizhou. All test results show that the technique of postponing nitrogen fertilizer has a higher yield, with a yield increase of 13.69%–23.17%. From 2005 to 2006, a comparison test was performed on double-cropping rice in Ganzhou, Jiangxi. The results show that compared with the rice subject to the "one-time fertilization" mode (10 : 0), the yield of rice under the technique of "postponing nitrogen fertilizer" (7 : 3) increased by 14.05% and 16.62% respectively in the early and late seasons.

⑤ Adjustment of the proportion of nitrogen fertilizer before and after organic basal fertilizer application. According to the results of the fixed-point test conducted by the Agricultural College of Yangzhou University, the proportion of nitrogen fertilizer 5.5 : 4.5 should be adjusted to 7 : 3 when all the straws are returned to the field, so as to increase the rapidly available nitrogen of basal fertilizer and make up for straw rot and nitrogen scramble among rice seedlings in the tillering stage. Nitrogen is released after straw decomposition, which is mainly used for spike fertilizer.

3) Reasonable nitrogen application technique.

① Application of base tiller fertilizer.

A. Proportion of base tiller fertilizer. Conventional rice basal fertilizer should generally account for 70%–80% of the total base tiller fertilizer, and tillering fertilizer should account for 20%–30% to reduce nitrogen loss. After transplanting, the fertilizer absorption ability of machine-transplanted seedlings is low, so the basal fertilizer should account for 20%–30% of the total base tiller fertilizer, and 70%–80% of which is concentrated in the new roots functioning as tillering fertilizer.

B. Time of application. Basal fertilizer is applied to the soil during land preparation, and part of it is used as foliar fertilizer. Tillering fertilizer should be applied as early as possible after the seedlings grow new roots, generally at the foliar age of 1st leaf (foliar age) after transplanting. As for machine-planted seedlings, tillering fertilizer should be concentratedly applied once or twice at the foliar age of 2nd and 3rd leaf after transplanting. Tillering fertilizer is generally applied only once. Fertiliser should not be applied in the middle and late tillering stages, so as not to cause vigorous growth in the ineffective tillering stage and so that the population cannot become "yellow" normally. In the case of insufficient population in the late tillering stage, it is better to make up through spike fertilizer, but not to supplement fertilizer in the late tillering stage.

② Accurate application and regulation of spike fertilizer.

A. The population is under normal seedling conditions. When there are sufficient seedlings at the critical foliar age for effective tillering (N–n or N–n+1), the leaf color begins to fade and become yellow, and the color of the top 4th leaf is lighter than the top 3rd leaf. The originally determined total amount of spike fertilizer can be applied twice as flower-promoting fertilizer (emergence of inverse 4th leaf) and flower-preserving fertilizer (emergence of inverse 2nd leaf). Flower-promoting fertilizer accounts for 60%–70% of the total amount of spike fertilizer, and flower-preserving fertilizer accounts for 30%–40%.

For the variety with four elongation internodes, the spike fertilizer should be applied once at the emergence of the inverse 3rd leaf.

When applying spike fertilizer, the water layer should not be kept in the field, and it is better to keep it in moist or shallow water. On the second day after application, the fertilizer is absorbed by the soil and then the shallow water layer should be irrigated, which will improve the fertilizer efficiency.

B. Insufficient population or early yellowing of leaves. There are not sufficient seedlings at the foliar age of N–n (for the variety with 4 elongation internodes, it is N–n+1), or the population becomes yellow early, which appeared at the foliar age of N–n (or N–n+1). In this case, the spike fertilizer should be applied to the variety with 5 elongation internodes in advance at the emergence of the inverse 5th leaf, and also at the emergence of the inverse 4th and 2nd leaf, a total of three times. The amount of nitrogen fertilizer applied should be increased by about 10% compared with the original plan, and the ratio of three times is 3 : 4 : 3.

In this case, the spike fertilizer should be applied to the variety with 4 elongation internodes in advance at the emergence of the inverse 4th leaf, and flower-preserving fertilizer should be applied at the emergence of the inverse 2nd leaf. The total amount of spike fertilizer can be increased by 5%–10%, and the ratio of flower-promoting fertilizer and flower-preserving fertilizer should be 7 : 3.

C. The population is too large and the leaf color is too dark. If the top 4th leaf is darker than the top 3rd leaf (top 4th > top 3rd) after the foliar age of N–n, the spike fertilizer must be postponed until the leaf color of the population becomes yellow, and the applied amount should be reduced as long as it is applied once.

Chapter IV Green Prevention and Control Technology for Diseases and Pests of Machine-transplanted Rice

Section I Occurrence and Identification of Main Diseases in Machine-transplanted Rice

I. Occurrence and Identification of Main Diseases in Rice

(Ⅰ) Seed-borne diseases

1. Rice bakanae disease [*Gibberella fujikuroi* (Saw.) Wollenw]

(1) Symptom identification.

Rice bakanae disease (Fig. 4-1) is a typical seed-borne disease that may occur from the seedling stage to the heading stage. Generally, there are three peaks of disease onset in the seedling stage, tillering stage, and jointing-booting stage. The diseased seedlings grow excessively and are yellowish and thin, obviously higher than about 1/3 of healthy seedlings, and many "anatropous fiber roots" grow on the nodes after jointing. Diseased seedlings begin to appear in the rice seedling bed. Dead seedlings may appear in the early stage of the field (tillering to jointing stage), a large number of dead seedlings may appear in the seriously diseased field, and pink mold (conidium of the pathogen) is produced on the culms and leaf sheaths. In some severely diseased fields, even seedlings in the whole field are destroyed. A small number of dead seedlings sometimes appear in the middle and late stages of the field, and there are often light-red powders and small black dots at the base, which is the perithecium of the pathogen. Although mildly diseased plants may head, the rice ears are small, the grains are few, or the rice ears are white.

Severely diseased grains turn brown and not plump, resulting in a mold layer on the glume.

Fig. 4-1 Rice Bakanae Disease

(2) Pathogen identification.

The teleomorph of the pathogen of bakanae disease is *Gibberella fujikuroi* (Saw.) Wollenw, which belongs to the *Gibberella* of Ascomycotina; the anamorph is *Fusarium moniliforme*, belonging to the *Fusarium* of Deuteromycotina. The pathogen is resistant to dryness. Mycelia can survive for 3 years on dry diseased straws, and conidia can survive for more than 2 years on dry diseased straws. The pathogen is not resistant to moisture, and mycelium dies in a short time on the wet soil surface or in soil.

(3) Infection cycle.

The pathogen is mainly attached to the seed surface and mycelium as conidium and incubated on the seeds or diseased straws for overwintering. The disease-free seeds are contaminated by bacteria-bearing seeds during seed soaking, and the pathogens infect the coleoptile during seed germination and result in disease infection, then produce conidia on the surfaces of diseased plants and reinfect it by airflow. During the flowering of rice, conidia fall on the stamens and pistils, germinate, and infect, causing the seeds to bear bacteria. Bacteria-bearing seeds are the primary source of infection for this disease. Wounds are conducive to pathogen infection, and the susceptibility among rice varieties varies greatly.

2. Rice white tip nematode disease (*Aphelenchoides besseyi* Christie) (Fig. 4-2)

(1) Symptom identification.

Rice white tip nematode disease (Fig. 4-2) is a typical seed-borne nematode disease that mainly damages leaves and rice ears, and dry tips appear when the seedlings have 4–5 leaves. It is most obvious after the booting stage, mostly distributed on the flag leave or at 1–8 cm of the tip of 2–3 leaves below it, showing a light yellow or yellowish-brown translucent appearance, and then twisting into a grayish-white dry tip, with one brown boundary line at the junction of diseased and healthy parts. Diseased plants generally can head normally, but are weak and short in growth, with short and narrow upper leaves, short rice ears, few grains, many blighted grains, and reduced thousand-grain weight. Some diseased plants do not show symptoms, but rice ears have nematodes, and seriously diseased fields may even form a large number of "small grains and erect ears".

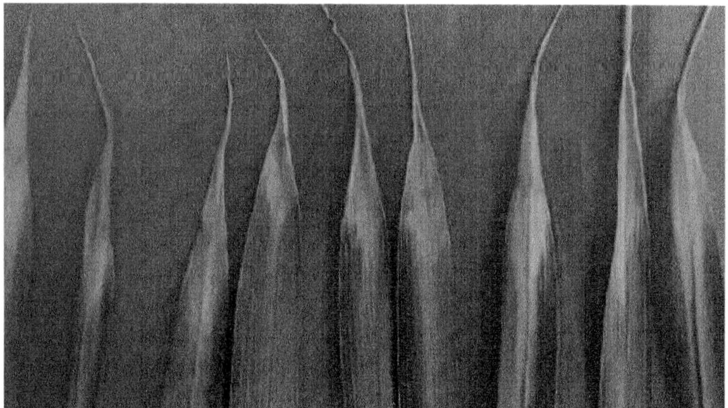

Fig. 4-2 Rice White Tip Nematode Disease

(2) Pathogen identification.

Pathogens of rice white tip nematode disease are *Aphelenchoides besseyi* Christie, which belong to the *Aphelenchoides* of the Aphelenchida.

(3) Infection cycle.

The degree to which this disease occurs is closely related to rice varieties and is mainly transmitted by nematodes in seeds. Once the disease occurs in the seedling bed or field, there is no treatment for such disease. The prevention and control method is mainly through seed treatment.

(Ⅱ) Fungal diseases

1. Rice sheath blight (*Rhizoctonia solani* Kühn)

(1) Symptom identification.

Water-stain-shaped dark green spots appear at positions near the water surface when the leaf sheath is infected, and the disease spots gradually expand into an oval or cloud shape (Fig. 4-3). When the conditions are suitable, the edges of the disease spots show dark green, and the center is grayish-brown, with a fast expansion speed. When the weather is dry, the edges are brown and the center is straw yellow to grayish-white. Due to tissue necrosis, the diseased leaf sheath can cause the leaves to turn withered and yellow. Disease spots on leaves are similar to those on leaf sheaths. When leaves are seriously diseased, it can cause the rice plant to be unable to head normally or even spread to the ear, resulting in an increase of shriveled grains and a decrease of grain weight, lodging, or the death of the whole plant. When the humidity is high, there are white to grayish-white arachnoid hyphae and oblate or irregular dark brown sclerotia in the diseased parts, and the hyphae are connected with sclerotia. In the later stage, a white-powdery mold layer can also be seen on the diseased part, which is the basidium and basidiospore of the pathogen.

Fig. 4-3 Rice Sheath Blight

(2) Features of damage caused.

Rice sheath blight may occur from both the seedling stage to the heading stage. The peak period is before and after the heading stage, which mainly damages leaf sheaths and leaves. In severe cases, it can infect the culms and spread to the rice ears. Sheath blight has two stages of disease onset in the lifetime of rice. The period from the peak tillering stage to the booting stage is the horizontal expansion stage, which expands horizontally between plants and holes, and the diseased hole rate and diseased plant rate increase. The period from the end of the booting stage to the waxy ripening stage is the vertical expansion stage, which spreads from the lower part to the upper part of rice plants, and the severity of the disease increases.

(3) Pathogen identification.

The anamorph of the pathogen of rice sheath blight is *Rhizoctonia solani* Kühn, which belongs to *Rhizoctonia* of Deuteromycotina; the teleomorph is *Thanatephorus cucumeris* (Frank) Donk., which belongs to *Thanatephorus* of Basidiomycotina. The pathogen development temperature is 10–36°C, and the suitable temperature is 28–32°C. Sclerotium germination requires a relative humidity of more than 96% and will be inhibited if the relative humidity is below 85%. Light can inhibit the growth of hyphae, but can promote the formation of sclerotia.

(4) Infection cycle.

The pathogen mainly overwinters as sclerotium in the soil, and can also overwinter as mycelium or sclerotium on diseased straws or other diseased residues. The number of sclerotium left in the field in the previous year is closely related to the disease severity in that year. After spring plowing and irrigation, the overwintering sclerotia float. After transplanting rice seedlings, the sclerotia float and attach to the leaf sheath at the base of the rice plant with water, germinate and grow hyphae at an appropriate temperature, extend from the leaf sheath surface into the inside to form appressoria, and then invade through stomata or directly through the epidermis. After the invasion, the sclerotia continue to expand in the rice plants and grow aerial hyphae outward, spreading to nearby leaves, leaf sheaths, or adjacent rice plants for secondary infection. The hyphae produced can also be spread by water flow, insects, or agricultural activities for secondary infection. In the late growth stage of rice, sclerotia formed on the surface of diseased parts are harvested with mature rice plants and fall off in the field for overwintering. After the sclerotia formed on the diseased part falls off, they can also float and attach to the base of the rice plant with water, causing secondary infection after germination.

2. Rice blast [*Pyricularia grisea* (Cooke) Sacc.]

(1) Symptom identification.

During the whole growth period of rice, it can cause damage to seedlings, leaves, leaf collars, nodes, ear necks, etc., which are called seedling blast, leaf blast, collar blast, node blast, neck blast, etc., respectively.

① Seedling blast and leaf blast (Fig. 4-4): the symptoms are similar and mainly divided into the following 4 types.

Chronic type: The disease spots are rhombic or spindle-shaped, with brown surroundings and grayish-white in the middle, with different sizes, up to 3–4 cm in length. A large number of such disease spots often indicate that weather conditions are not conducive to disease onset, and a grayish-green mold layer is produced in the diseased part under a humid environment.

Acute type: The disease spots are mostly oval or irregular, water-stain-shaped in dark green, and the diseased parts are covered with grayish-green mold layers. The high incidence of such disease spots indicates a tendency for leaf blast to be prevalent.

Brown-spot type: A few needle-shaped and small disease spots in brown, which mostly appear on disease-resistant varieties and old leaves at the lower part of the plant.

White-spot type: The disease spots are suborbicular small white spots. They mostly appear in dry, sunny, and hot weather and are changeful during the stage of symptomatic appearance and very unstable. In the case of unsuitable conditions after the disease spots appear, they can convert into chronic-type disease spots. If conditions are suitable, they can convert into acute-type disease spots.

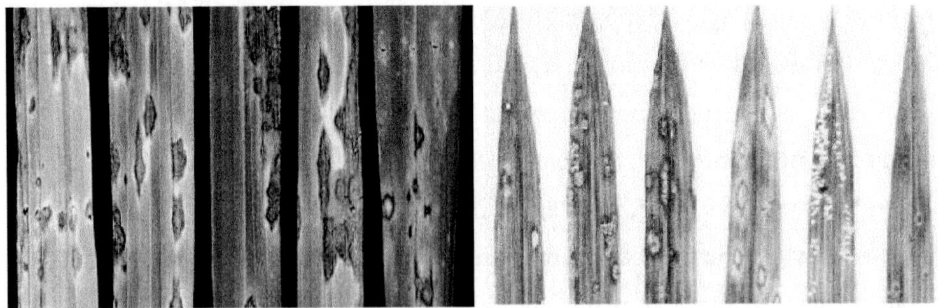

Fig. 4-4 Leaf Blast

② Collar blast (Fig. 4-5) is a general term for the disease of ligule, auricle, and pulvinus in the leaf collar part. The diseased part is dirty-green initially and grayish-brown after expansion, often causing early withering of leaves or the occurrence of neck blast.

Fig. 4-5 Collar Blast

③ Node blast (Fig. 4-6) mostly occurs at 1–2 nodes below the ear. The disease spots are brown and small initially, and then expand to the whole node in a ring shape, in dark brown. In high humidity, there is a large amount of gray mold layer produced on the diseased parts. In the later stage, the diseased node dries and shrinks and is easy to break, resulting in early withering above the diseased node.

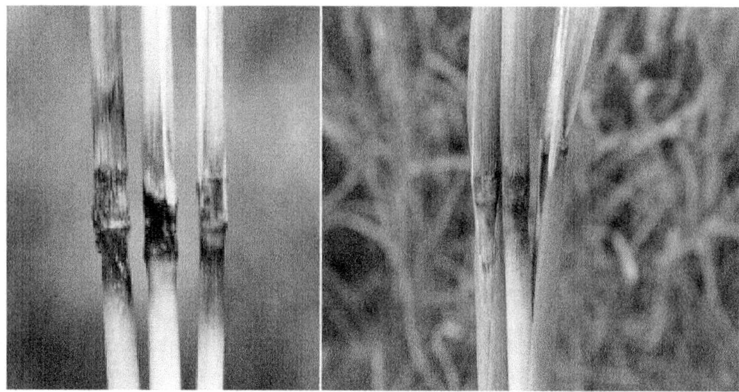

Fig. 4-6 Node Blast

④ Neck blast (Fig. 4-7) occurs from the main pedicel to the neck of the first branch. The diseased part is small water-stain-shaped brown dots initially, then gradually expand and turn brown or dark green. In the later stage, the central part of the diseased part is withered and white, and easy to break. This disease can also develop on rachides and branches. A gray mold layer can be produced on all diseased parts. White and grizzled ears are mostly formed in rice with early onset of the disease, while rice with late onset of the disease has more shriveled grains and lower grain weight, affecting the quality of rice. In addition, the pathogen can also infect grains and empty glumes.

Fig. 4-7 Neck Blast

(2) Pathogen identification.

The pathogenic bacteria of rice blast pathogen belong to the *Pyricularia* of Deuteromycotina. The teleomorph is Ascomycotina and the anamorph is Deuteromycotina. The temperature for hypha germination is 15–32℃ and the optimal temperature is 25–28℃. The germination requires a relative humidity of more than 90%, preferably with water droplets or water films.

(3) Infection cycle.

The pathogens mainly overwinter on diseased grains and straws, and become the source of primary infection in the spring of next year. Conidia can survive for half a year to one year, and hyphae in diseased tissues can survive for more than one year. The overwintering pathogens in grains invade the plumule during seed soaking and pregermination, causing primary infection, necrosis, and rot of seedlings. The overwintering pathogens on the diseased straws begin to produce conidia from late May to early June, which scatter with airflows and attach to the seedlings. Under suitable conditions, they cause primary infection. In a humid environment, the diseased part of the primary infection produces conidia, which are transmitted to healthy plants by airflows, causing frequent reinfection. The suitable temperature range for pathogen invasion is 20–32°C, and the optimal temperature is about 24°C. The pathogen does not invade below 13℃ or above 35℃. Where susceptible varieties are planted in a large area, it is easy to cause disease. Climatic conditions have a direct impact on the infection by pathogens and the disease resistance of rice plants. The seedling blast and leaf blast occur in the 3–5 leaf stages and tillering stages of rice. Generally, when the temperature is 17–32°C, especially 24–28°C, the relative humidity reaches 90%, and foggy, dewy, rainy, or humid weather occurs, such as in the East Asian rainy season, the seedling blast and leaf blast are prone to occur due to this type of weather. If there is a significant temperature decline during the spike initial appearance stage and heading stage of single-cropping and post-cropping rice, the stress resistance of rice will decrease. If there is much rainfall or foggy and dewy weather, it is conducive to the occurrence of the neck blast. The management of cultivation fertilizer and water significantly affects the occurrence degree of rice blast. The heavy application of nitrogen fertilizer or too late application of earing fertilizer will lead to the vigorous growth of rice plants, unfavorable-delayed senescence in the later stage and delayed heading, aggravating the occurrence degree of leaf blast and neck blast.

3. Rice false smut [*Ustilaginoidea virens* (Cke.) Tak.]

(1) Symptom identification of damage features.

Rice false smut (Fig. 4-8) occurs after the flowering to the milky stage. After the pathogen invades the grain, it first forms an enlarged sporodochidium to wrap the whole grain, which is yellowish-green to green. Then, the surface cracks and is covered with dark green powder, which is the chlamydospore of the pathogen. Some sclerotia can be formed in the sporodochidium of diseased grains and appear on the surface of diseased grains when chlamydospores are scattered and fall off after maturity.

The occurrence of rice false smut not only affects the rice yield and reduces the setting rate and thousand-grain weight, but the pathogen also contains toxins harmful to humans and animals, which attach to the grains and pollute the rice crop, seriously affecting the quality.

Fig. 4-8 Rice False Smut

(2) Pathogen identification.

The anamorph of the pathogen is *Ustilaginoidea virens* (Cke.) Tak., which belongs to *Chlorobium* of imperfect fungi; and the teleomorph is *Claviceps oryzaesativae* Has., which belongs to *Claviceps* of Ascomycota. In addition to infecting rice, the pathogen can also infect maize, medicinal wild rice and other plants. The optimal temperature for the growth of hyphae is 28–30°C, and the growth rate is obviously slowed down when the temperature is lower than 13°C or higher than 35°C. The optimal temperature for sclerotium germination and production of the fruiting body is 26–28℃.

(3) Infection cycle.

The pathogens overwinter as chlamydospores or sclerotia in the soil or attached to seeds, and produce conidia and ascospores in the summer and autumn of the next year, which are transmitted by airflow and damage floral organs and young glumes, and are the main primary infection sources. Conidia produced by chlamydospores play an important role in secondary infection. Disease symptoms are generally seen 10–12 days after heading; the peak of disease onset is in 20–25 days and becomes basically stable in 30–35 days.

(Ⅲ) Bacterial diseases

1. Rice bacterial leaf blight [*Xanthomonas oryzae* pv. *oryzae* (Ishiyama) Swings et al.]

(1) Symptom identification.

Rice bacterial leaf blight (Fig. 4-9) mostly occurs after the tillering stage and mainly damages leaves. The symptoms vary with rice varieties, period of disease onset and infection parts. It is usually divided into four types: common type, acute type, withering type, and yellow-leaf type. The common type first produces yellowish-green or dark green spots on the leaf tip or leaf margin, then expands up and down along the leaf margin or midrib to form streaks, and the diseased tissue is grayish-white after withering, which is the most common symptom of the leaf blight type. In the case of abundant fertilizer, tender green plants, rainy and sultry weather, and extremely susceptible varieties, the whole leaves lose water and curl rapidly after the onset of disease, showing a withering state in dark green like scalded by boiling water, which is an acute symptom. The appearance of such symptoms indicates that the disease is developing rapidly. In the southern rice area, some susceptible varieties may show withering symptoms in the case of a large number of pathogens or injured rhizomes. In high humidity, the surface of the diseased part with various symptoms often overflows dew-like yellow pus colloids (called bacterial pus), which form roe-like small colloidal particles after drying and are easy to fall off.

Fig. 4-9 Rice Bacterial Leaf Blight

(2) Infection cycle.

Bacteria-bearing seeds, diseased straws, and residual diseased plants in the fields are the main sources of primary infection. Field weeds such as *Leersia hexandra* Swartz. also spread the disease. Bacteria overwinter in seeds and invade through water holes and wounds on leaves after sowing. The central diseased plant is formed, and the diseased plant secretes yellow bacterial pus, which is transmitted by wind and rain, dew, insects, human factors, and other factors. Through irrigation water, wind, and rain, pathogens spread over long distances, low-lying accumulated water, waterlogging, and flood irrigation may cause continuous disease. Operation in fields still with the morning dew may cause the spread of pathogens. High temperature and humidity, dew, typhoon, and heavy rain are the conditions of prevalence of diseases. Long-term water accumulation, excessive nitrogen fertilizer, excessive growth, and acidic soil in paddy fields are all conducive to the occurrence of diseases. Rice is susceptible to the disease at the young panicle differentiation stage and booting stage.

2. Rice bacterial leaf streak [*Xanthomonas oryzae* pv. *oryzicola* (Fang et al.) Swings et al.]

(1) Symptom identification.

Rice bacterial leaf streak (Fig. 4-10) mainly damages rice leaves. The disease spots are small dark green translucent spots in the early stage and then expand along the leaf vein to form dark green or yellowish-brown slender streaks, which are oil-stained-shaped and translucent when observed to the light. In severe cases, the disease spots grow together and make the whole leaves wither and yellow. When the humidity is high, dew-like yellow bacterial pus appears on the disease spots and is aligned in rows. This disease may occur from the seedling stage to the heading stage, and mostly occurs from the end of the tillering stage to the early stage of heading. Indica rice is extremely susceptible to disease in general, and most japonica rice has stronger resistance.

Fig. 4-10 Rice Bacterial Leaf Streak

(2) Infection cycle.

In the Yangtze River basin, most of the pathogens overwinter on seeds and diseased grasses, becoming the source of primary infection in the coming year and an important way of long-distance transmission. The pathogen invades through the stomata and wounds of rice leaves, spreads closely through rainwater, irrigation water, insects, and farming operations, and spreads through contact between leaves. The bacterial pus on the surface of diseased leaves is the source of reinfection in the field. Storms during the rice growing season are the main cause of pathogen spread in fields. The suitable temperature for disease onset is 25–30°C. Under suitable temperature conditions, the greater the rainfall and the higher the field humidity, the earlier and more severe the disease infection.

3. Rice bacterial foot rot (*Erwinia chrysanthemi* pv. *zeae*)

(1) Symptom identification.

Water-soaked elliptical spots are often produced on the leaf sheaths at the base of the culm near the soil surface and gradually expand into irregular large spots with brown edges and withered white in the center (Fig. 4-11). When the leaf sheaths are peeled off, the root and nodes turn black and brown, sometimes dark brown longitudinal strips can be seen. The root and nodes rot, accompanied by a stench, and the inner leaves of the plant wither and turn yellow. When rice is infected in the tillering stage, the diseased plants first show green and curly inner leaves, then gradually wither and turn yellow and look like withered seedlings caused by borers. The disease often suddenly outbreaks during rainstorms or in flooded fields, causing disease in the whole field and a large number of dead seedlings. At the jointing stage of round culm, the leaves of diseased plants gradually turn yellow from bottom to top, and the leaf sheaths near the water surface show long strip-shaped disease spots with brown edges and caesious in the center. When rice is infected after the booting stage, the symptoms are often manifested as acute wilting and seedling death. The diseased plants lose water and wither, forming withered booting, semi-withered panicles and withered panicles. Some of the 2–3 culm nodes above the base of the diseased plants also turn black and brown at the same time, with a small number of anatropous roots.

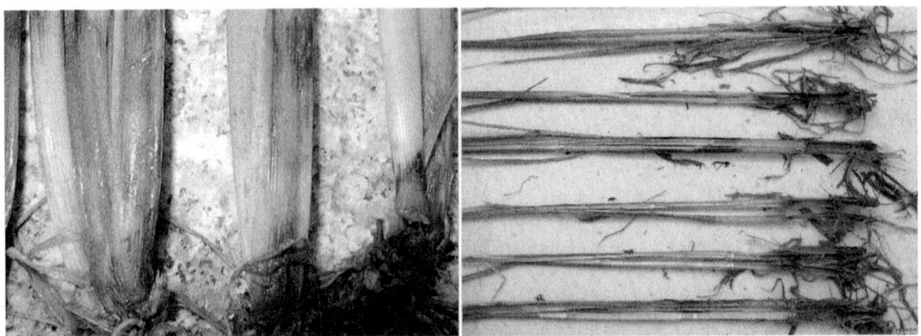

Fig. 4-11 Rice Bacterial Foot Rot

(2) Infection cycle.

Bacteria can overwinter on diseased straws, diseased rice stubbles, and weeds. The pathogen invades from the water holes and wounds on the leaves, leaf sheaths, and the root system, mainly from the root or wounds on the culm base. After the invasion, the pathogen causes infection systematically in stomata at the root base and causes repeated infections during the whole growth period. Early rice begins to show symptoms after transplanting and enters the peak of disease onset in the heading stage. The disease often occurs in late rice seedling beds and it reaches its peak in the booting stage. The disease is mild in rice with crop rotation, direct sowing, or seedling transplanting. Heavy application or late application of nitrogen makes rice seedlings tender and leads to severe disease infection. The disease is prone to occurring at the end of the tillering stage without dehydration or excessive soil drying. The disease is prone to being severe in low terrain and clayey soil, which is poorly aerated. Generally, late rice has earlier disease onset than early rice.

(Ⅳ) Rice virus disease

1. Rice stripe virus (RSV)

(1) Symptom identification.

Rice can be damaged by RSV (Fig. 4-12) transmission by *Laodelphax striatellus* (Fallen) during the seedling stage, from the tillering stage to the jointing stage and booting stage and show symptoms. The most susceptible growth period is from the seedling stage to the tillering stage. In the early stage, the diseased plants first appear with irregular chlorotic streaks or yellowish-white stripes parallel to the leaf veins on the inner leaves (seedling stage) and then merge into large pieces. Half or most of the diseased leaves turn yellowish-white, and then the newly-grown inner leaves gradually turn yellow and curl, forming a bending and drooping "falsely withered inner leaves" in a paper twist shape. Some diseased plants of japonica rice show redness of old leaves in the middle and late stages of the disease. When plants show symptoms and become diseased in the seedling stage, it often leads to withering and death; when plants become diseased in the tillering stage, the tillers of diseased plants often decrease, and most of the seriously diseased plants die as a whole. When plants become diseased in the heading stage, the diseased ears are characterized by deformity and unfruitfulness.

Fig. 4-12 Rice Stripe Virus (RSV)

(2) Pathogen identification.

The pathogen of the rice stripe virus is RSV, which belongs to *Tenuivirus*. The virion is circular or filamentous, and the viral nucleic acids are four single-stranded RNA, which form a branched filamentous or rod-like hyperstructure. The virus is mainly transmitted by the *Laodelphax striatellus* (Fallen). *Laodelphax striatellus* (Fallen) generally takes 30 minutes to suck the diseased plants to be infected with the virus (only 3–10 minutes for a few individuals). After it is infected with the virus, it requires a circulative period for virus transmission. The circulative period is 4–23 days, with an average of 8.3 days. After the circulative period, the virus can be transmitted for 40 days. The virus is transmitted through eggs, and there is still a high virus transmission rate till the 40th generation.

(3) Infection cycle.

Rice stripe virus mainly overwinters in the nymphs of *Laodelphax striatellus* (Fallen), and some overwinter in diseased plants of barley, wheat, and weed, which is the source of primary infection of the disease in the next year. From late March to mid-April, the nymphs of overwintering *Laodelphax striatellus* (Fallen) migrate to wheat fields after eclosion, lay eggs and reproduce in wheat fields and weeds of Poaceae and part of them invade the nearby seedling beds and cause damage. After the eclosion of the first-generation imagoes from late May to early June, as the wheat ripens and harvests, a large number of them migrate into the seedling beds and early sowing and direct sowing paddy fields to cause damage and spread viruses. After 10–15 days, the first peak of symptoms appears in fields of early-planted middle and late rice, early sowing and direct sowing paddy fields. Subsequently, with the alternation of generations and eclosion of imagoes of *Laodelphax striatellus* (Fallen), the second and third peaks of virus transmission and symptomatic appearance occur. The nymphs of the sixth generation of *Laodelphax striatellus* (Fallen) hatch in early and mid-October, and after the maturity of rice, the virus-carrying nymphs of *Laodelphax striatellus* (Fallen) migrate into field weeds of Poaceae and wheat fields to cause damage and overwinter.

2. Rice black-streaked dwarf virus (RBSDV)

(1) Symptom identification.

RBSDV (Fig. 4-13) is mainly characterized by dwarfing of diseased plants, dark green and stiff leaves, short irregular strip-shaped protrusions in waxy white at the early stage and in dark brown at the later stage along the leaf vein on the surface due to cell hyperplasia of the phloem on the back of the leaf, leaf sheath, and culm. The tillers of diseased plants increase, but this phenomenon does not exist in seedlings with early onset. Rice plants with early onset are obviously dwarfed and cannot head; those with late-onset have small ears and poor fruiting. There are rounded inclusion bodies in the proliferative histiocytes of the diseased plant, with a diameter of 6.5 μm.

Fig. 4-13 Rice Black-streaked Dwarf Virus (RBSDV)

The onset of disease mostly begins before and after heading and some plants have early onset before and after the peak tillering stage of rice. When rice plants have early onset, the diseased plants are dwarfed and their leaves turn yellow. Yellowing occurs mostly in the middle leaves of the diseased plant (2–3 leaves above and below the third leaf centered below the sprout leaves). The yellow color is more obvious from the middle of the leaf to the tip, while the tip turns brown; sometimes the leaf turns orange, often with brown necrotic spots. The diseased leaves with mild symptoms in the upper part have small spots of discoloration, but generally, the newly sprouted leaves and the next leaves are normal without discoloration. When the disease onset occurs later before and after heading, the flag leaf and its next superior leaf are only obviously dwarfed, but the appearance and leaf color are normal, just prone to turn yellow earlier. The yellowing diseased plants are dwarfed, but they are also sometimes dwarfed after the yellowing symptoms appear. Generally, the length of internodes, leaf sheaths, and leaves from the 4th to 5th leaf positions below the flag leaf of the diseased plant is shortened. There are also slightly narrow and twisted leaves and rachides turn brown, and brown streaks are produced on glumes. The heading stage is roughly the same as or slightly later than that of healthy plants, and the diseased plants rarely wither and die, but also sometimes die when common dwarf disease concurs. Plants with disease onset before and after heading are 12%–30% shorter than the healthy plants and impurities, immature rice, and deformed rice are increased, with poor gloss. The appearance, taste, fragrance, and hardness of rice deteriorate.

The root system is underdeveloped, with a small number of fine roots, sparse and hard, poor elasticity, slightly turning brown, a small number of upper roots, poor viability, and early withering and death. Due to insufficient water absorption, the leaves are prone to curl under strong sunlight. Symptoms are characterized by yellowing and dwarfing during the onset of the early stage and the onset in the later stage is characterized by dwarfism. Seen from a distance, the diseased plants in the field are dwarfed and in clusters, the leaves turn pale and appear to lack fertilizer, and the yellow leaves are obviously scattered among them. After the application of earing fertilizer, the leaf color of healthy plants becomes thick, and the yellow color of diseased plants is obvious. This is because the diseased plants absorb fertilizer poorly and the effect of fertilizer is difficult to show, while most of the diseased clusters in the diseased field are concentrated to form clustered diseased nests, and there are also scattered diseased clusters distributed in the field. There are rounded, elliptical, strip-shaped diseased nests of different sizes, scattered in the field or at the edge of the field. In severe cases, the whole field is in a concave-convex shape. However, the field sometimes can be low on the whole and a few healthy plants can be seen scattered in the field. In the center of the clustered diseased nest, the whole cluster is dwarfed, and some diseased plants are scattered around it, including the whole-cluster dwarfed plants and partially dwarfed ones. Dwarf-diseased plants in clusters are easy to be identified and scattered diseased plants are difficult to find. There are two types of diseased nests: the dish type, characterized by a horizontal ground surface and the conical type, which has a distinct center and a periphery that gradually changes from short to high. The former may be acute, with symptoms occurring simultaneously in a short period and the latter may be chronic, with symptoms expanding slowly over a longer period.

(2) Pathogen identification.

The virions in the diseased plants and virus-carrying pests are spherical, with a diameter of 80–90 nm and a diameter of 60 nm in the purified sample. When determining with the micro-injection method, the dilution endpoint for the sap of diseased leaves is 10^{-4}–10^{-5}, that for the sap extracted from virus-carrying pests is 10^{-5}–10^{-6}, and the passivated temperature of the sap of diseased leaves is 50–60℃ for 10 minutes. The survival time *in vitro* at 4℃ is 6 days and the pathogen can still remain highly infectious at -30– -35°C after 232 days.

(3) Infection cycle.

RBSDV is a type of virus disease. The virus transmission vectors include *Laodelphax striatellus* (Fallen) and *Sogatella furcifera* (Horváth) and the virus is mainly transmitted by *Laodelphax striatellus* (Fallen). Once the vector is infected, it carries the virus for life, but the virus is not transmitted through eggs. The virus mainly overwinters on diseased plants of barley and wheat, and can also overwinter in the body of *Laodelphax striatellus* (Fallen). Viruses in fields complete the infection cycle through the path of wheat–early rice–late rice.

Laodelphax striatellus (Fallen) cannot transmit the virus to offspring through eggs and can only carry the virus after sucking on pathogenic plants such as dwarfed rice and dwarfed wheat. Therefore, RBSDV is transmitted by repeated transfer of *Laodelphax striatellus* (Fallen) on rice and wheat. *Laodelphax striatellus* (Fallen) occurs 5–6 generations a year in the suburbs of Shanghai, China, and overwinter as the third to fourth instar nymphs of the fifth and sixth generations. Virus-carrying overwintering *Laodelphax striatellus* (Fallen) spread viruses from late rice to wheat, causing dwarfed wheat; in the second year, the first generation of *Laodelphax striatellus* (Fallen) obtain viruses from dwarfed wheat, and migrate from wheat to early rice from mid- to late May, resulting in dwarfed rice; in late June and mid- to late July, the second and third generations of *Laodelphax striatellus* (Fallen) migrate from early rice to late rice seedling beds and fields, respectively, spread the disease and causing damage; later, the overwintering *Laodelphax striatellus* (Fallen) on the late rice transmit viruses to the wheat again, and repeat this process for numerous times.

3. South rice black-streaked dwarf virus (SRBSDV) (Fig. 4-14)

(1) Symptom identification.

The main symptoms of rice plants infected with SRBSDV include increased tillers, short, wide, and stiff leaves, and dark green leaves. The veins and culms on the back of the leaves show waxy white in the early stage, and turn brown in the later stage, with short-striped knot-like protuberances, no heading or with small ears, poor fruiting, and symptoms slightly vary after infection at different growth stages. When the disease occurs in the seedling stage, the inner leaves grow slowly, and leaves are short, wide, stiff, and dark green, with irregular waxy-white knot-like protuberances on the leaf vein, which then turn blackish-brown. The roots are short and the plants are dwarfed, without heading, and often wither and die early. When the disease occurs in the tillering stage, the newly-grown tiller shows symptoms first. Short and small diseased ears are still sprouted from the main stem and early tiller, but the diseased ears are hidden in the leaf sheath. When the disease occurs in the jointing stage, the flag leaf is short and broad and the ear neck is shortened, with a low setting rate. There are short strip-shaped nodular protuberances on the back of leaves and culms. After the onset of the disease, the typical features of rice are dwarfed plants, dark green leaves, and strip-shaped milky white or waxy white appear on the leaf back and culms and then turn to dark-brown small protrusions subsequently, high tiller and anatropous fiber roots on the culm nodes, no heading or small ears, and poor fruiting. There are uneven wrinkles on the flag leaf or upper leaves, and one or several rice plants in one cluster are about 1/3 shorter than the healthy plants, with semi-ears or complete ears.

① Typical symptoms.

The leaves of the diseased rice plants are dark green, and uneven wrinkles are seen on the leaf surface of the upper leaves (mostly at the base of the leaves). There are anatropous fiber roots and high nodal branches at the nodes of several nodes above the ground of the diseased plant. The surface of the culm of the diseased plant has milky white knot-like protuberances about 1–2 mm in size (with obvious rough feeling when touched by hand), and knot-like protuberances are aligned in strips longitudinally in the shape of wax dots and are milky white in the early stage and brownish-black in the later stage. The nodal region produced by disease nodules varies with different infection periods. If the rice plant is infected in the early stage, the disease nodules are on the lower node, and the later the infection period is, the higher the nodal region produced by disease nodules is.

Fig. 4-14 South Rice Black-streaked Dwarf Virus (SRBSDV)

② Symptoms in the seedling stage.

The color of the diseased plant is dark green, the inner leaves grow slowly, and the leaves are short, small, stiff, and dark green, with irregular waxy-white knot-like protuberances on the leaf vein, which then turn blackish-brown. The spacing between the pulvinus is shortened, and its leaf sheath is wrapped in the lower leaf sheath. The plant is short and small (less than 1/3 of the normal plant) and cannot head in the later stage, often withering and dying early.

③ Symptoms in the tillering stage.

When the diseased plants have more tillers and clusters, the upper leaves have overlapping pulvini, the inner leaves break the lower leaf sheath and emerge, or stretch out spirally from the pulvinus mouth of the lower leaf, the leaves will grow short and stiff, and the leaf tip is slightly twisted and deformed. Although the plant is short, the main stem and early-grown tiller can still head, but it is difficult for the ear head to bear fruit, or the ear is wrapped, or the ear is small, similar to dwarfism.

④ Symptoms in the heading stage.

The whole plant is dwarf and clustered, some of which can head, but it is relatively late and small in heading, with few filled grains and light grain weight. It is half wrapped in the leaf sheath, and the flag leaf is short and rigid; longitudinal folds can be seen at the base of the middle and upper leaves; waxy white or blackish-brown raised short-striped vein swelling can be seen at the lower internodes, and nodes of the culm.

(2) Pathogen identification.

SRBSDV belongs to the Fiji disease virus (FDV) of Reoviridae. Observation by electron microscope shows that SRBSDV has a morphology similar to that of FDV of the same genus, i.e., the virions are icosahedral spherical, with a diameter of about 66–70 nm, non-enveloped, composed of double coats, and have 12 "A-spike"-shaped protrusions with a length and a diameter of about 11 nm at the vertex angle of the icosahedron. The diameter of the inner core is about 55 nm, and there are 12 "B-spike"-shaped protrusions with a length of about 8 nm and a diameter of about 12 nm in the inner core. SRBSDV virions are composed of 10 double-stranded dsRNA, which are named S1–S10 according to the mobility on gel electrophoresis SDS-PAGE. *Sogatella furcifera* (Horváth) is the main vector of virus transmission. In addition to rice, the virus also damages more than 20 kinds of grain crops and weeds, such as maize, wheat, *Digitaria sanguinalis* (L.) Scop., *Alopecurus aequalis* Sobol. , *Echinochloa crusgalli* (L.) Beauv., etc.

(3) Infection cycle.

The virus is mainly transmitted by *Sogatella furcifera* (Horváth), which does not transmit the virus through eggs. The time for obtaining the virus by *Sogatella furcifera* (Horváth) is 30 seconds and the virus transmission time is 15 seconds. The virus cannot be transmitted through seeds, and plants do not transmit the virus to each other. Once the vector is infected, it carries the virus for life, and the incubation period of rice plants after inoculation is 14–24 days. The virus-carrying *Sogatella furcifera* (Horváth) that migrates from other places is the main source of primary infection of the virus, and the virus-carrying host (such as regenerate seedlings and weeds in the field) can also become the source of primary infection after winter; the virus can be transmitted by the virus-carrying (virus-obtaining) *Sogatella furcifera* (Horváth) feeding on the host plant.

II. Occurrence and Identification of Main Pests in Rice

(I) Pests in the rice seedling stage

1. Rice thrip [*Stenchaetothrips biformis* (Bagnall)]

(1) Morphological identification.

① Imago: 1–1.3 mm long, dark brown, with a head approximately square and 7 segments of antennae; light-brown wings, feather-shaped; females have cone-shaped abdomen ends, and males have relatively round and blunt abdomen ends (Fig. 4-15).

Fig. 4-15 Imagoes of Rice Thrip

② Egg: kidney-shaped, about 0.26 mm long, yellowish-white.

③ Nymph: 4 instars in total. The 4th instar nymph is also called pupa, with a length of 0.8–1.3 mm, light yellow, and antennae folded to the back of the head and chest.

(2) Identification of damage features (Fig. 4-16).

Imagoes and nymphs rasp and break the leaf surface with their mouthparts, causing tiny yellowish-white spots. Both sides of the leaf tip roll and fold inward, and gradually the whole leaves curl, wither, and turn yellow. In paddy fields that are seriously damaged at the early tillering stage, the seedlings and roots do not grow and no tillers, and plants even wither and die in clusters. Seedling beds of late rice are more severely damaged and often die in patches, looking like burned by fire. Imagoes and nymphs at the heading stage tend to damage earbuds, and during flowering, they turn into glumes, which damage the ovary and cause empty and shriveled grains.

Fig. 4-16 Damage Features Caused by Rice Thrips

(3) Occurrence regularity.

Rice thrips have a short life cycle, many generations, and overlapping generations. Most of them overwinter as imagoes in wheat fields, *Zizania latifolia* (Griseb.) Turcz. ex Stapf, and weeds of Poaceae. Imagoes often hide in the rolled leaf tip or inside the inner leaves and are active in the morning and evening and on cloudy days. They have an obvious tendency to oviposit on tender green rice seedlings and the eggs are oviposited scattered between the leaf veins. After the formation of young ears, they mostly oviposit on the inner leaves. The newly hatched larvae are concentrated in auricle and ligule, and prefer to damage young inner leaves. In July and August, it is low-temperature and rainy, which is conducive to damage. The seedling stage, tillering stage, and young panicle differentiation stage are the severe damage stages of thrips, especially in the late rice seedling beds and the early transplanting fields.

2. *Laodelphax striatellus* (Fallen)

(1) Morphological identification.

①Imago: two types, namely long-winged and short-winged. The long-winged imago is 3.5–4.0 mm long, and the short-winged imago is 2.4–2.6 mm long; the vertex is square, which protrudes slightly in front of the compound eyes; the middle of the frons is the widest, and the lateral carina is curved. The body color of males is dark brown and the body color of females is yellowish-brown. The areas between lateral carinae on the anterior half of the vertex, the face, and the thoracal pleura are all dark brown. The posterior half of the vertex, the pronotum, the mesothorax tegula, the carinae of the frons and the clypeus, the antennae, and the legs are all light yellowish-brown. The mesonotum of males is dark brown, while that of females is light yellow, with dark brown wide stripes on both sides. The abdomen of males is dark brown, the back of the abdomen of females is dark brown, and the ventral surface is yellowish brown. The forewings are light yellowish-brown and transparent, the veins are the same color as the wing surfaces, and the wing spots are dark brown (Fig. 4-17).

② Egg: banana-shaped, about 0.8 mm long and 0.2 mm wide. The bottom width of the egg cap is greater than the height, and the top is blunt and round. The eggs are exposed on the egg-laying marks in bead shape.

③ Nymph: mostly experiences 5 instars. It is long oval shaped and milky yellow or orangish-yellow at the 1st and 2nd instars, with a light-colored longitudinal band in the middle of the thorax; it is milky white or light yellow at the 3rd to 5th

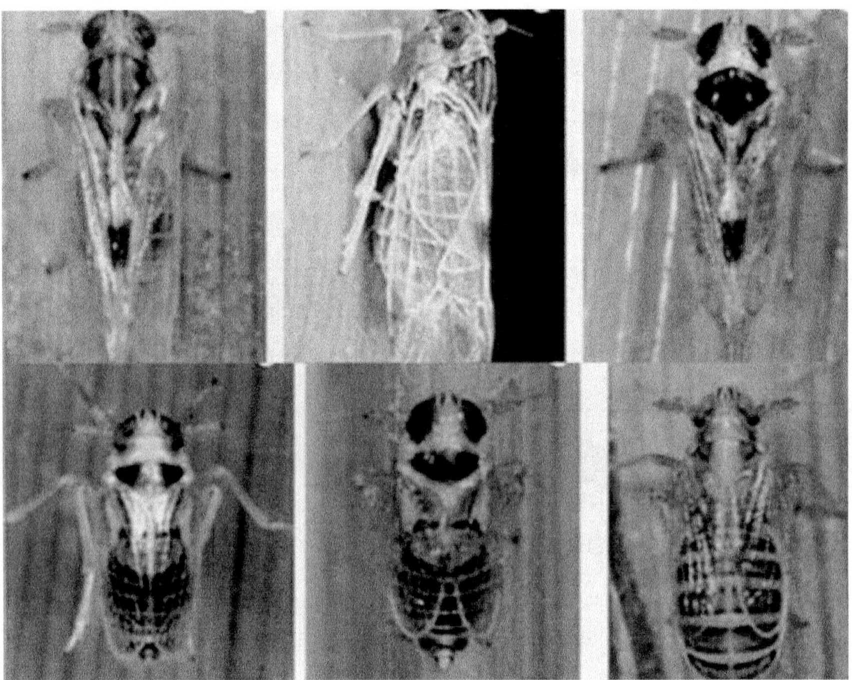

Fig. 4-17 Imagoes of *Laodelphax striatellus* (Fallen)

instars, the longitudinal band in the middle of the thorax turns milky yellow, with brown patterns on both sides and reversed-V-Shaped light-colored patterns on the back of the third and fourth abdominal segments, the abdomen end is relatively blunt and round, the wing buds are obvious, and the legs extend backward obliquely in a reversed-V shape after falling into the water (Fig. 4-18).

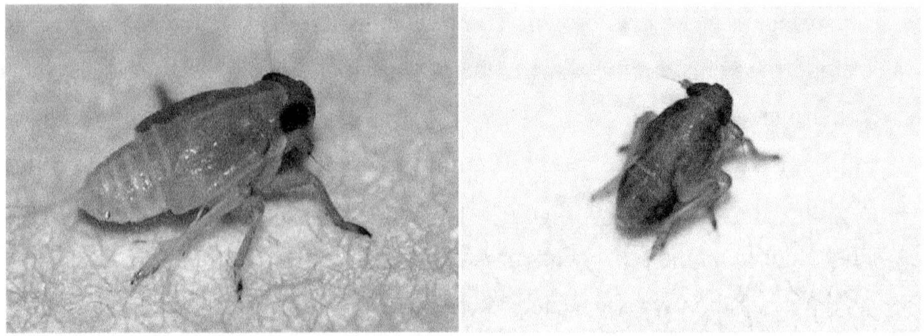

Fig. 4-18 Nymphs of *Laodelphax striatellus* (Fallen)

(2) Identification of damage features.

In spite of the damage of *Laodelphax striatellus* (Fallen) imagoes and nymphs to rice by piercing and sucking, the *Laodelphax striatellus* (Fallen) will spread rice stripe virus (RSV), rice black-streaked dwarf virus (RBSDV), wheat rosette virus (WRV), and maize rough dwarf virus (MRDV), and the loss caused by virus transmission of *Laodelphax striatellus* (Fallen) is far greater than that caused by direct sucking. Sooty blotch symptoms caused by insect feces honeydew are common at the base of rice bushes with a high density of *Laodelphax striatellus* (Fallen). In a few years, sooty blotch will even occur at the spike.

(3) Occurrence regularity.

Five to six generations occur every year in Jiangsu Province, which most overwinter as nymphs on "three kinds of wheat" (wheat, barley, hull-less barley) and gramineous weeds such as *Alopecurus aequalis* Sobol and *Polypogon fugax* in green manure fields. From late May to early June, nymphs become the first-generation imagoes by eclosion, and when the "three wheats" mature, the *Laodelphax striatellus* (Fallen) migrates into the middle and late rice seedling beds and the fields of early and middle rice to oviposit, reproduce, and damage plants.

(4) Biological characteristics.

① Living habits: *Laodelphax striatellus* (Fallen) can overwinter locally, spread in small areas as long-winged imagoes, and show signs of long-distance migration. *Laodelphax striatellus* (Fallen) has the habits of phototaxis, green-taxis, and edge-taxis. The population density of *Laodelphax striatellus* (Fallen) on the edge is much higher than that in the fields. Imagoes in paddy fields can oviposit on barnyard grass, and the number of eggs on barnyard grass is often 5–11 times higher than that on rice. The overwintering females oviposit the most eggs, and the eggs are mostly oviposited in the lower leaf sheath of rice plants and the tissues on both sides of the middle vein of leaf bases.

② Virus transmission characteristics: Virus transmission by *Laodelphax striatellus* (Fallen) has different characteristics with different virus species. After sucking the sap of diseased plants with RSV, the *Laodelphax striatellus* (Fallen) carries the virus for life and the virus can be inherited by generations; it will still have strong virus transmission ability after 40 generations. After sucking the sap of diseased plants with RBSDV, the *Laodelphax striatellus* (Fallen) imagoes and nymphs can also carry the virus for life, and the virus can overwinter in the nymphs but cannot be transmitted to the next generation with eggs. The characteristics of transmitting MRDV by *Laodelphax striatellus* (Fallen) are similar to that of transmitting RBSDV and WRV, which cannot be transmitted through eggs. Virus transmission of virus-carrying *Laodelphax striatellus* (Fallen) has the characteristics of intermittent transmission. The viability of affected *Laodelphax striatellus* (Fallen) is reduced, hence its life span is shortened, the number of eggs is reduced, the hatching rate is reduced, and the mortality rate is increased.

(Ⅱ) Rice migratory pests

1. *Sogatella furcifera* (Horváth)

(1) Morphological identification.

① Imago: The long-winged imago is 3.8–4.6 mm long (including the length of the wing) and the short-winged is 2.5–3.5 mm long. The vertex is rectangular, prominently protruding in front of the compound eyes; the lateral carina on the frons is straight, with the widest part at 1/3 of the end. Most of the males are dark brown, and most of the females are light yellowish-brown. The vertex (except for the area between the lateral carinae on the apical part), the pronotum, and the middle area of the mesonotum are yellowish-white. There is a dark brown spot behind the compound eyes on the pronotum, the lateral area of the mesonotum is dark brown, and the same of females is slightly light. For males, the area between the lateral carinae on the apical part of the vertex and the face is dark brown; for females, this area is yellowish-brown, and only the abdomen and back have blackish-brown spots; the forewings are light yellowish-brown and transparent, sometimes with smoke brown halo at the end of the wings, and the wing spots are blackish-brown (Fig. 4-19).

② Egg: 0.8 mm long and 0.2 mm wide, crescent-shaped. Eggs are milky white at birth, then turn yellow with red eye spots. The egg cap has a height greater than the bottom width, tapering toward the end, and is not exposed or slightly exposed in the egg-laying marks.

Fig. 4-19 Imagoes of *Sogatella furcifera* (Horváth)

③ Nymph: five instars in total. It is nearly olive-shaped, with pointed ends. After falling into the water, the nymph extends its hindlegs flat to both sides into a "—" shape.

(2) Identification of damage features.

Imagoes and nymphs cluster at the base of the rice bushes to suck the sap from stem and leaf tissues and consume the nutrients of rice plants. *Sogatella furcifera* (Horváth) mainly damages rice plants before the heading of rice, and the damage effects manifest as short plants, short spikes, small spikes, and reduced fruiting rate. When severely damaged, the orangish-yellow plants turn gradually into sauce-brown and remain upright without collapse. However, in a few years when *Sogatella furcifera* (Horváth) occurs frequently, severely damaged indica rice will appear the phenomenon of withering and lodging. In the later stage, when the number of pests is large, they will feed on the ears, resulting in discoloration of the chaffs, half-shrunk of grains, and excrete honeydew, which will lead to sooty mold and blackening of the ears.

(3) Occurrence regularity.

It usually starts to migrate into Jiangsu around June 20, and the migration peak in southern Jiangsu and along the Yangtze River starts from June 25 to the end of June. There are 1–3 immigration peaks in July, generally one peak earlier than *Nilaparvata lugens* (Stål). Usually, the offspring of the first immigrant imagoes is the main damage generation. In the year or region when the number of immigrants is very large, the immigrant imagoes will cause obvious damage and the phenomenon of "causing disaster at the moment of arrival". In the one-season late rice fields, only early immigrants become short-winged imagoes by eclosion. In general, most of the fifth (2) generation of imagoes migrate outward after eclosion, and the population number drops sharply. In individual years, most of the fifth (2) generation stay and convert into the sixth (3) generation to continue to cause damage.

(4) Biological characteristics.

Imagoes have phototaxis and migration. There are two types of long-winged and short-winged, and the proportion of long-winged in each generation is always more than 80%. A certain number of short-winged females appear when the population density is low, rainfall is abundant and the nutritional conditions of rice are good. Short-winged males are rare. Their eggs are mostly oviposited on the back of leaves. Nymphs inhabit and feed at the base of rice plants, and the position selected by them is higher than that by *Nilaparvata lugens* (Stål). After the milk ripening of rice, they often migrate to the main veins and spikes of sword leaves and feed on such places.

2. *Nilaparvata lugens* (Stål)

(1) Morphological identification.

① Imago: two types, namely long-winged and short-winged. Long-winged imagoes are 3.6–4.8 mm (including the length of the wing), and there are two types of body color, namely dark and light. The dark-colored body is brown to dark brown, with oily luster, and the vertex is nearly square, which is brown as the pronotum and mesonotum; the frons is nearly rectangular, the middle part is slightly wide, the face and antennae are light dark brown; the forewings are yellowish-brown and transparent, and the wing spots are dark brown. Female imagoes are large and light in color, while male imagoes are small and dark in color. The light-colored body is brown, and only the thorax, abdomen, and back of the abdomen are dark. The short-winged females are 3–4 mm long, with an enlarged abdomen, a blunt and round abdomen end, and forewings extending to 5–6 segments of the abdomen. The short-winged males are 2.5 mm long, and most of the forewings extend to the 5th segment of the abdomen; the hindwings are all degenerated. The paramere of males is in the shape of crab claws, with the end protruding inward and anteriorly in a sharp angle shape; the base of the inner margin of the first valvifer of females is protruding in a semicircular shape (Fig. 4-20).

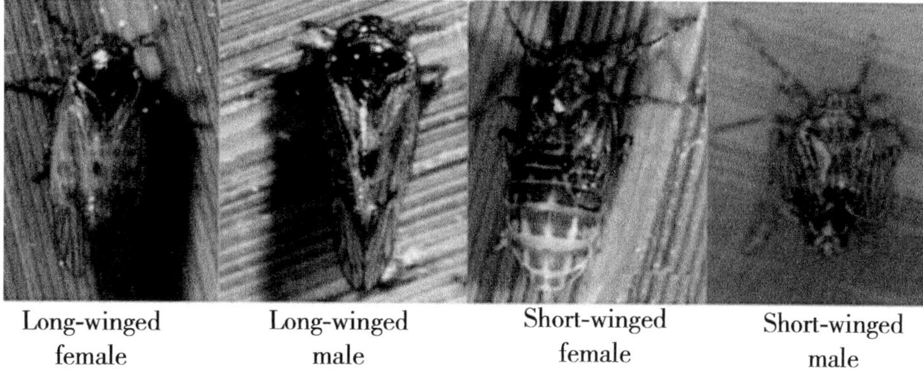

Fig. 4-20 Imagoes of *Nilaparvata lugens* (Stål)

② Egg: banana-shaped, about 1 mm long and 0.22 mm wide, slightly curved, the height of the egg cap larger than the bottom width, with an arc-shaped top and slightly exposed egg-laying marks. The exposed part is nearly short oval, and looks like small squares, clear and countable. Eggs are laid in veins and leaf tissues and arranged in a line, called "egg strips". It is milky white at birth, gradually turns light yellow to rusty brown, and red eye spots appear in the later stage.

③ Nymph: five instars in total, ovoid, with two types of body color, namely light and dark, and the color difference is greater after the 3rd instar. The body of dark-colored is yellowish-brown, with dark brown stripes on the back of the abdomen, the body of light-colored is light yellow, and the stripes on the back of the body are not obvious (Fig. 4-21).

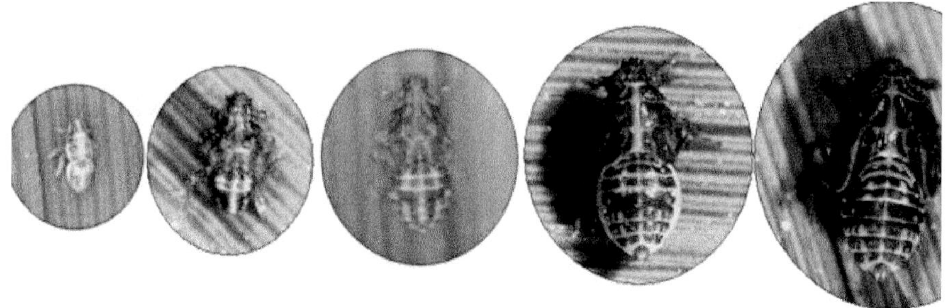

Fig. 4-21 Nymphs of *Nilaparvata lugens* (Stål)

(2) Identification of damage features (Fig. 4-22).

The imagoes and nymphs cluster at the base of the rice bushes, directly piercing and sucking the sap from stem and leaf tissues, consuming the nutrients of rice plants, making the grains become hollow, increasing the rate of empty seed, and decreasing the thousand-grain weight. Since the coagulative saliva secreted during piercing and sap-sucking hinders the nutrient and water transmission of rice plants, when the number of pests is large and rice plants are seriously damaged, the lower part of the rice plant will become black, rot, and stink, and the plant lodging will occur, which may lead to serious yield reduction or no harvest. In addition to direct damage, it can also spread WRV and RDV, wounds caused by feeding and egg-laying facilitate the infection of sheath blight and culm rot, and excreted honeydew easily induces the occurrence of sooty mold.

Fig. 4-22 Damage Features Caused by *Nilaparvata lugens* (Stål)

(3) Occurrence regularity.

Nilaparvata lugens (Stål) is a thermophilous pest that cannot overwinter in Jiangsu. Every year, the first batch of immigrant pests in Jiangsu comes from the South (mainly in the Nanling area and its south) along with the warm and humid airflow for a long distance. In the year of major occurrence, it shows an obvious "two peaks" migration. *Nilaparvata lugens* (Stål) can complete three generations in the rice areas in the south of the Yangtze River and the rice areas along the Yangtze River in Jiangsu. The first immigrant long-winged imagoes are the fourth (1) generation, and the offspring bred through egg-laying is the fifth (2) generation.

When conditions are suitable, there will be more short-winged imagoes in the fifth (2) generation after eclosion, and the reproduced sixth (3) generation is the main damage generation. From the immigrant generation to the production generation and then to the main damage generation, there is an alternate-generation outbreak. Most of the sixth (3) generation of imagoes upon eclosion in the middle-season rice areas in the north of the Huahe River and Yangtze-Huaihe River basins are long-winged and immigrate outward. The wing types of imagoes of different generations have obvious alternations of length. The sixth (3) generation imagoes in the southern late-season rice areas are partially inhabited to produce the seventh (4) generation to keep damaging rice plants, and the seventh (4) generation occurs more frequently in warm autumns.

(4) Biological characteristics.

There are two wing types of *Nilaparvata lugens* (Stål): long-winded and short-winded. Short-winged imagoes are mostly produced when the population density is small, the nitrogen content in rice is high, and the light and humidity conditions are suitable. The short-winged imagoes have a short pre-oviposition period, a long life span, a high amount of oviposition, and are not good at flying, belonging to the residence type. The increase in short-winged imagoes indicates that the population is about to proliferate in large numbers. Long-winged imagoes have the characteristics of phototaxis and strong flying ability. They are diffusible and migratory and have the habit of seasonal long-distance migration. The eggs are laid in strips in the midrib of the leaf sheath or in the midrib tissue near the base of rice leaves. The newly hatched nymphs mostly concentrate on the rice straw about 3 cm above the water surface for feeding. With the increase in the number and instars of pests, the feeding site gradually rises and even to the top of the rice plant. *Nilaparvata lugens* (Stål) has many kinds of natural enemies. Among them, insects that have significant inhibitory effects on *Nilaparvata lugens* (Stål) include *Anagrus nilaparvatae*, which parasitize eggs; *Cyrtorhinus lividipennis* (Reuter), which sucks egg fluid; as well as Dryinidae, *Pipunculus* sp. , nematodes, and *Beauveria,* which parasitize nymphs and imagoes. The predatory natural enemies are mainly spiders and rove beetles.

(Ⅲ) *Cnaphalocrocis medinalis* (Guenée)

(1) Morphological identification.

① Imago: The imago is 7–9 mm long. Wing span is 12–18 mm. It is yellowish-brown, with blackish-brown wide stripes on the outer margins of the forewings and hindwings, and the costal margins of forewings are dark brown, with 3 blackish-brown transverse lines; the middle transverse lines are short and do not extend to the trailing margins. There are 2 blackish-brown transverse lines on the hindwings. The male moth is small, with an eye-shaped stripe composed of a black hair cluster in the center of the forewing's costal margins, and a cluster of black hairs at the base of the forelegs' tarsus. When it is at rest, the tail often tilts upward. The female moth is large and has a straight tail at rest (Fig. 4-23).

Fig. 4-23 Imagoes of *Cnaphalocrocis medinalis* (Guenée)

② Egg: The egg is suborbicular. It is about 1 mm long and 0.5 mm wide, flat, slightly raised in the center, with reticulated stripes on the surface of eggshells. It is milky white at birth and light yellow before incubation. The parasited eggs are dark brown.

③ Larva: The mature larva is 14–19 mm long, generally with 5 instars and a few with 6 instars. The body color changes from light yellowish-green to yellowish-green and turns orange when it is nearly mature. At the 1st instar, the head is black, and the body is thin and 1–2 mm long. After the 2nd instar, the head changes from yellowish-brown to brown, and the brown stripes on the pronotum and the hairs on the back of the mesothorax and metathorax are gradually visible. At the 4th instar, brown stripes on the pronotum and the black stripes around the hairs on the back of the mesothorax and metathorax are in the darkest color (Fig. 4-24).

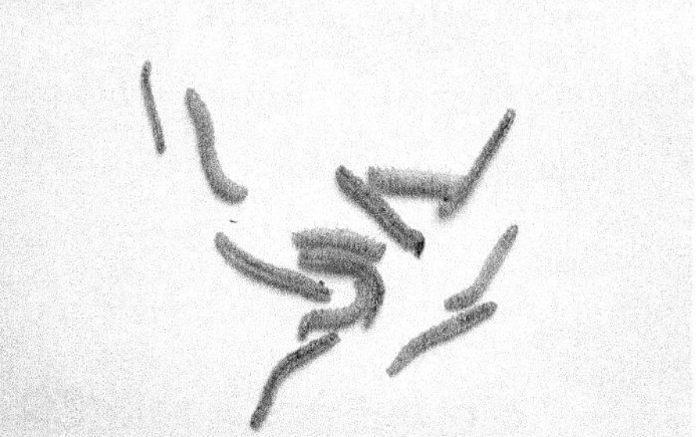

Fig. 4-24 Larvae of *Cnaphalocrocis medinalis* (Guenée)

④ Pupa: The pupa is 7–10 mm long, oblong cylindrical, with pointed and thin ends. The buttock spine is obviously protruding with 8 hook spines, and the trailing margin of the back of each abdominal segment is raised. It is light yellow at first and then turns reddish-brown to brown (Fig. 4-25).

Fig. 4-25 Pupae of *Cnaphalocrocis medinalis* (Guenée)

(2) Identification of damage features (Fig. 4-26).

The newly hatched larvae inhabit the inner leaves or young leaves of late tillering and gnaw at mesophyll. In the booting stage, they inhabit the ear buds and gnaw the chaffs and leaf sheaths, resulting in translucent white thin strip-shaped spots. In the 2nd instar, they spin silk at the tip of the inner leaves or rice leaves, and the margins on both sides of the leaves are connected to form rice leaf bundles, which is called the leaf-bundling stage. In the 3rd instar, form monotubular bracts, leaving a white epidermis after the mesophyll is eaten up. The feeding amount at 1st–3rd instars accounts for about 20% of the total feeding amount in a lifetime. After the 3rd instar, the larvae enter the overeating stage and often bundle 2–4 leaves to form a bract, and the larvae feed in the bracts, which results in a large number of white leaves. In the late stage of rice growth, the number of empty grains increases and the grain weight decreases due to the damage to functional leaves.

Fig. 4-26 Damage Features Caused by *Cnaphalocrocis medinalis* (Guenée)

(3) Occurrence regularity.

The annual occurrence number of *Cnaphalocrocis medinalis* (Guenée) in Jiangsu is 2–3 generations, with the fourth (2) and fifth (3) generations as the main damage generations. *Cnaphalocrocis medinalis* (Guenée) is weak in cold resistance and has no habit of low-temperature diapause. In the area north of the isotherm with an average temperature of 4°C in January (equivalent to 30°N), no insect stage can overwinter. Every year, the first batch of immigrant pests in Jiangsu is mainly from the northern part of the Lingnan area, and the imagoes migrate to Jiangsu with the southwest airflow in summer. At the areas in front of the trough in frontal weather, migratory imagoes land along with rainfall or downdraft (imago landing may also occur at the edge of subtropical anticyclone). The sixth (4) generation of imagoes migrates outward in large numbers. In warm autumn, some imagoes stay on late-maturing single-season late rice to lay eggs and damage rice plants.

(4) Biological characteristics.

Imagoes have characteristics of migratory and strong flying ability. The booting period of imago eclosion is after 20:00, and the peak appears around midnight. Mating occurs 1–2 days after eclosion and is mostly carried out in the second half of the night, with the peak occurring from 3:00 to 5:00. Imagoes mostly hide in dense shade crops or weeds during the day and fly for a short distance when they are disturbed. Imago has phototaxis, especially to metal halide lamps, and prefers to suck nectar from plants and honeydew from aphids as supplemental nutrition. In the same growth period, the number of eggs laid in the green rice field is several times or even more than ten times higher than that in ordinary fields. The oviposition sites are mostly on the back of the middle and upper leaves of the plant, especially the basipetal 2nd and 3rd leaves. The newly hatched larvae generally crawl into the inner leaves or the nearby leaf sheaths first, and there are also some that may get into the old bracts to eat mesophyll, forming needle-sized white transparent dots. At the 2nd instar, larvae begin to spin and roll longitudinally into small bracts at the tip of the leaf during the leaf-bundling stage. After the 3rd instar, they begin to move to other bracts for damage. The food intake increases sharply from the 4th to the 5th instar. Larvae can damage 5–7 leaves and up to 9–10 leaves in a lifetime. The larvae are very active and fall down quickly when the curled leaves are peeled off. The mature larvae spin silk and make thin cocoons to pupate after 1–2 days of the prepupa stage.

If nitrogen fertilizer is applied too much or applied too late, the rice seedlings will grow excessively with drooping leaves, the nitrogen content in the plant will be high with more eggs laid, and the damage will be severe. The growth and development of *Cnaphalocrocis medinalis* (Guenée) require moderate temperature and high humidity, with an appropriate temperature of 22–28°C and a relative humidity of more than 80%. There are more than 80 natural enemies that prey on or parasitize *Cnaphalocrocis medinalis* (Guenée) in each insect stage. The parasitic natural enemies in the egg stage include *Trichogramma confusum*, *Trichogramma japonicun* Ashmead and *Trichogramma dendrolimi*. In the larval stage, there are rice leaf folder parasitoids, rice bollworm parasitoids, and flatted wasps. In the pupal stage, there are Ichneumon, *Brachymeria lasus* Walker, and parasitic flies. Predatory natural enemies include frogs, dragonflies, damselflies, carabids, rove beetles, and spiders. Among them, the common *Trichogramma confusum*, rice leaf folder parasitoids, and *Erigonidium graminicola* have great control effects on *Cnaphalocrocis medinalis* (Guenée).

(Ⅳ) Rice boring pest

1. *Sesamia inferens* Walker

(1) Morphological identification.

① Imago: Imagoes are 12–15 mm long, 27–30 mm wide, with light yellowish-brown cephalothorax, light yellow abdomen, rectangular light brownish-yellow forewings, and distinct dark brown longitudinal line in the middle of the wings (from the base of the wings, even to the outer margins), with two small black spots above and below this line. The hindwings are silver-white, the antennae of male moths are pectiniform, and the antennae of female moths are filamentous (Fig. 4-27).

Fig. 4-27 Imagoes of *Sesamia inferens* Walker

② Eggs: Eggs are oblate, slightly concave at the top, 0.5 mm in diameter and 0.3 mm in height, with radial thin carinas on the surface, egg mass in a band shape, arranged in 2–4 rows. Eggs are milky white at birth and then turn light yellow, light red, to black (Fig. 4-28).

Fig. 4-28 Eggs of *Sesamia inferens* Walker

③ Larva: Larvae are divided into 6 instars. The mature larva is about 30 mm long, and thick, with a reddish-brown head, a light yellow trunk, and a purplish-red back. There are 12–15 crochets of abdominal legs arranged in a median band (Fig. 4-29).

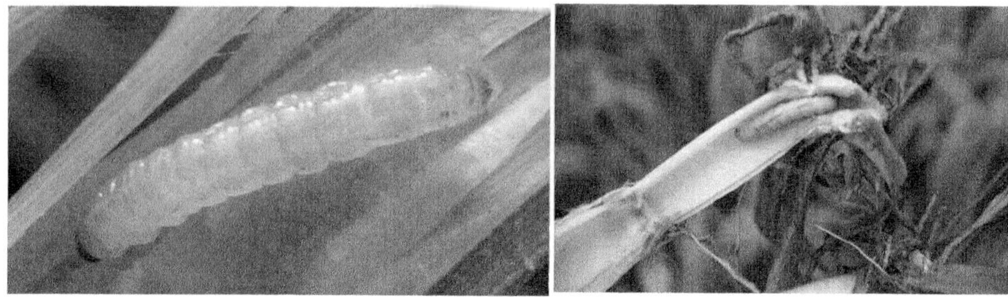

Fig. 4-29 Larvae of *Sesamia inferens* Walker

④ Pupa: Pupae are 15–18 mm long, milky white at the initial stage, and gradually turning yellowish-brown and brown. When the eclosion is about to take place, the body is reddish-black, with white powder attached to the head, and there is a section of the left and right wing buds connected, with legs not extending out of the wing buds, and there are 4 protrusions at the end of the abdomen (Fig. 4-30).

(2) Identification of damage features.

The newly hatched larvae in paddy fields first cluster in leaf sheaths to feed, resulting in withered sheaths, and scatter to damage stems after 2nd–3rd instars, resulting in a large number of withered seedlings. After that, the larvae continue

Fig. 4-30 Pupae of *Sesamia inferens* Walker

to transfer to other plants for damage. About 20 days after incubation, the seedling withering stops developing. The eggs are laid in the ear buds at the booting stage, then the eggs hatch into larvae. The larvae first damage the young ears in the ear buds, causing withered ears. After heading, the larvae crawl out and get into the ear neck to form white ears. For eggs in the leaf sheath, the larvae first feed intensively after hatching, and then invade the rice stem after the second instar. For rice plants that have headed, the larvae mostly directly drill holes from the sword leaf sheath into the ear neck, causing white ears and insect-bitten plants. When the larvae enter the late instar, they begin to transfer downward and bore into rice stems, causing insect-bitten plants.

(3) Occurrence regularity.

Three generations of *Sesamia inferens* Walker occur in a year in Jiangsu, and the fourth generation completely occurs in some areas in warm autumn. However, in areas dominated by hybrid rice, 4 generations may occur every year. After the larvae overwinter, the first-generation imagoes begin in mid- to late April and are prevalent in early to mid-May, with 2–3 moth peaks in general and 4–5 moth peaks at most in some years. In areas where rice and spring maize coexist, most of the first-generation imagoes lay eggs in spring maize fields, causing withered inner leaves and insect-bitten plants. The second-generation imagoes initially occur from late June to early July, and fully occur in early and mid-July. The larvae damage middle rice and premature late rice from late July to early August, causing withered inner leaves. The third-generation imagoes initially occur from late July to early August, and fully occur in early and mid-August. The larvae damage the middle-late rice and post-cropping rice from late August to early September, causing white ears, insect-bitten plants, and withered ears.

(4) Biological characteristics.

Imago eclosion mostly occurs from 19:00 to 20:00. They inhabit weeds or the base of rice bushes during the day and move from 20:00 to 21:00. The peak period for them to flutter around lights is from 23:00 to 1:30 the next day. The first-generation imagoes have strong phototaxis, the second and third generations have weak phototaxis at high temperatures, and the fourth generation has stronger phototaxis. The sex pheromone of the first-generation male moths has a strong attractive effect. The second and third generations have a poor attractive effect, so attraction by lamps or sex pheromones often cannot reflect the growth and decline of imagoes in high-temperature seasons. Imagoes of *Sesamia inferens* Walker prefer to oviposit in open spaces. In paddy fields with high density, eggs are mostly laid on the edge of the field. As the density decreases, the eggs of *Sesamia inferens* Walker gradually expand to the field center. Female moths are highly selective in oviposition. The first-generation imagoes prefer to oviposit on the inner side of the second leaf sheath at the base of corn. The third-generation imagoes prefer to oviposit on barnyard grass at the field edge. There are also a few egg masses distributed on rice plants at the field edge with tall stems, thick stems, and dense leaf sheaths. The eggs of the third generation are mostly laid on the rice seedlings at the booting stage and full heading stage. The eggs are mainly laid on the inner side of the rice leaf sheath, mostly within 3 cm from the pulvinus. The surface of the leaf sheath with eggs bulges, with faded egg marks that seem to have been scalded by boiling water.

In the mixed planting zone of spring maize and rice, the correlation between spring maize area and population proliferation of *Sesamia inferens* Walker was proved. Because there are many water bamboos and reeds in lakeside areas, the food for *Sesamia inferens* Walker is increased. The mixed planting of middle and late rice or the expansion of the rice area (triple-cropping rice planted after other plants) is beneficial to the increase of the population of *Sesamia inferens* Walker. The mixing rate of barnyard grass in the field is high, which is conducive to the oviposition of *Sesamia inferens* Walker. The common natural enemies of the *Sesamia inferens* Walker include predatory natural enemies such as frogs, ducks, and ground beetles, as well as Braconidae, Ichneumonidae, and parasitic flies that parasitize larvae and pupae.

2. *Chilo suppressalis* (Walker)

(1) Morphological identification.

① Imago: The male moth has a body length of 10–12 mm, a wing span of 20–25 mm, and a yellowish-brown cephalothorax. It has a nearly rectangular, yellowish-brown or grayish-brown forewing, with scattered brown spots on the wing surface, a purplish-black spot in the center, 3 small spots of the same color arranged obliquely below it, and 7 black spots on the outer margin. Its hindwings are white, gradually yellowish-brown near the outer margin, and the abdomen is thin. The female moth has a body length of 12–15 mm and a wing span of 25–31 mm. The head, thorax, and forewings are yellowish-brown, with a few brown spots on the surface of the forewings and 7 small black spots on the outer margin, and without purplish-black spots. The moth has white and silky hindwings and a fat abdomen (Fig. 4-31).

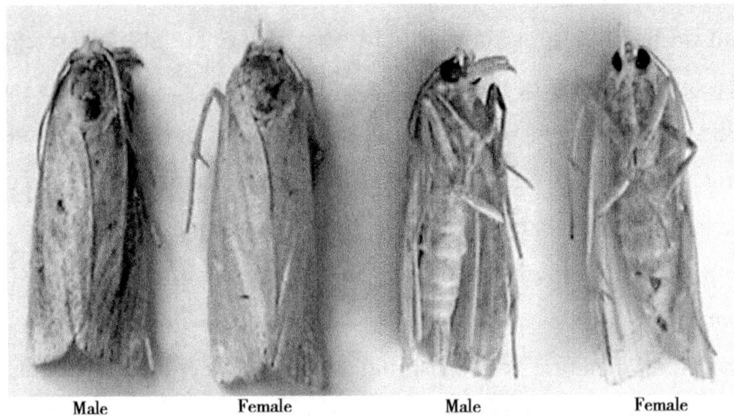

Fig. 4-31 Imagoes of *Chilo suppressalis* (Walker)

② Egg: The eggs are flat oval, 1.0 mm×0.7 mm in size, with inconspicuous reticulated stripes. They are white at birth, pink afterwards, then grayish-brown to grayish-black, generally with tens to hundreds of eggs arranged in fish scales and long strips (Fig. 4-32).

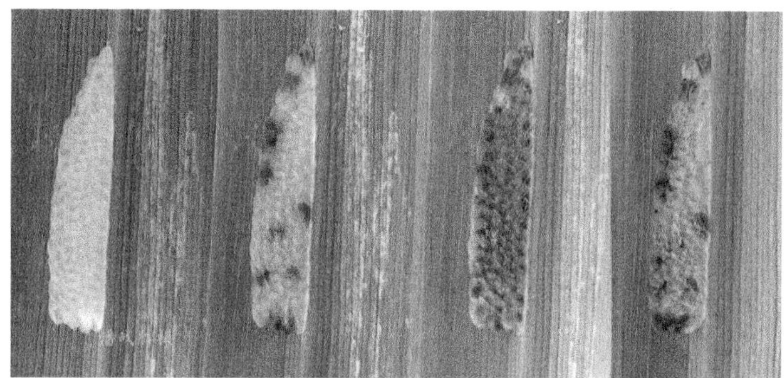

Fig. 4-32 Eggs of *Chilo suppressalis* (Walker)

③ Larva: The growing larvae are 20–25 mm long, light brown, with 5 reddish-brown longitudinal lines on the back, and the lowest one passing through the spiracles; the crochets of abdominal legs are biordinal without circles. They usually have 6 instars (Fig. 4-33).

Fig. 4-33 Larvae of *Chilo suppressalis* (Walker)

④ Pupa: It is about 13 mm long, yellowish-brown or brown, with 5 brown longitudinal lines faintly visible on the back of the abdomen. The head has papillae protruding to the abdomen. The proboscis is 1/2 of the wing length, and the end of the median leg is flat with the wing tip. Pupation occurs in stems of rice and water bamboo or between stems and leaf sheaths (Fig. 4-34).

Fig. 4-34 Pupae of *Chilo suppressalis* (Walker)

(2) Identification of damage features (Fig. 4-35).

The first-generation imagoes mainly lay eggs on the seedlings of middle and late rice, which will be transplanted to the field with eggs or young larvae. The field of early-planted middle rice may also be damaged by the first-generation larvae, causing withered sheaths and withered inner leaves. The second-generation larvae damage the middle and late rice and the triple-cropping rice planted after other plants (after-melon rice, after-bean rice, and after-corn rice) in the field, causing withered sheaths, withered inner leaves, withered ears, white ears, and insect-bitten plants. The insect-bitten plants in the later stage of the field with high population density may collapse in clusters, resulting in lodging.

Fig. 4-35 Damage Features Caused by *Chilo suppressalis* (Walker)

(3) Occurrence regularity.

Two generations occur all year round in Jiangsu, and incomplete third generations occur in some years with high temperatures. The second-generation mature larvae overwinter in rice roots, and a few overwinter in straws. In spring, the young overwintering larvae can transfer to the stems of "three kinds of wheat" and oilseed rape to get nutrients, resulting in white ears or withered ears of wheat. The farming system has an obvious influence on the occurrence of *Chilo suppressalis* (Walker). If the farming system tends to be simple, the population of *Chilo suppressalis* (Walker) will increase accordingly. A variety of *Trichogramma* and *Telenomus dignus*, as well as a variety of Ichneumonidae, parasitic flies and nematodes, *Beauveria bassiana*, etc., have a higher parasitic rate of *Chilo suppressalis* (Walker) eggs and a greater effect on inhibiting the occurrence of *Chilo suppressalis* (Walker).

(4) Biological characteristics.

Imago eclosion mostly occurs at night. Imagoes lie quietly in the paddy fields and weeds during the day and fly out at night. They have strong phototaxis and are more sensitive to black light. Imagoes copulate on the night of eclosion or the next day, and begin to lay eggs one day later. Most of the female moths lay eggs in fields with high plants, thick stalks, long sword leaves, and dark green leaves, and the number of laid eggs is the largest in tillering and booting stages of rice. After the hatched borer comes out of the shell, most of them crawl downward along the rice leaves or spin and droop, invading through the gaps or boreholes of the inner leaves and leaf sheaths. When the seedlings are young, the hatched borers are mostly scattered in small clusters. After 5–6 foliar ages, the hatched borers first concentrate on the inner side of the leaf sheath to feed on the inner wall tissue of the leaf sheath. The damaged leaf sheath changes color after 2–3 days, and withers and turns yellow after 7–10 days, i.e., the withered sheath stage. At this time, the larvae have not yet invaded the inner leaves or rice stems, which is a favorable time for investigation and control. After the second instar, the larvae begin to bore into rice stems, forming withered inner leaves, withered ears, white ears, and insect-bitten plants. It is usually seen that several or tens of larvae damage one plant. If the food is insufficient, the larvae will transfer to other plants for damage. When the growth of rice plants is blocked due to dry weather and water shortage, the larvae transfer frequently and the damage is aggravated. After maturing, the larvae pupate on the inner side of the stem or leaf sheath, and the overwintering larvae pupate inside the stem of rice piles, straw, or summer-ripe crops. The pupation part in the rice straw often rises and falls according to the water level.

3. *Scirpophaga incertulas* (Walker)

(1) Morphological identification.

①Imago: The male moth is 8–9 mm long, with a wing span of 18–22 mm, and its head, back of the thorax, and forewings are light grayish-brown. The compound eyes are grayish-black, and the labial palp is long, extending forward. The forewings are nearly triangular, with a small black spot in the center (lower corner of the mid cell), a dark brown oblique band from the wing tip to the center of the inner margin, and 9 small black spots on the outer margin (the front 7 are more obvious). The marginal hair has the same color as the wings. The hindwings are grayish-white, with light brown on the margin, with 3 anal veins, and white marginal hair. The male moth has a thin and grayish-white abdomen and a 3-segment yellowish-brown base. The female moth is 10–13 mm long, with a wing span of 23–28 mm, and the body is yellowish-white or light yellow. The forewings are light yellow or yellow, darker outward, with a distinct black spot at the lower corner of the mid-cell, and yellow marginal hair. The end of the abdomen has light dark-brown or yellowish-brown hair, which falls off after oviposition (Fig. 4-36).

Fig. 4-36 Imagoes of *Scirpophaga incertulas* (Walker)

② Egg: The eggs are 0.3–0.9 mm long and 0.7–0.8 mm wide, flat, and oval, but they are often irregularly quadrilateral or pentagonal due to being squeezed together. They are milky white at birth, then turn yellowish-white and yellowish-brown, and turn grayish-black when they are nearly hatched. The egg mass is 6.3 mm long and 2.8 mm wide on average, oblong, with a slight bulge in the center, and shaped like a half-sided soybean. The egg grains are arranged in three layers in the center, with one or two layers at the edge. The surface of the egg mass is covered with light dark brown or light yellowish-brown scale hairs, which are arranged in disorder.

③ Larva: The larvae generally have 4–5 instars. Mature larvae are about 21 mm long, and their heads are light brown or yellowish-brown. The thorax and abdomen are yellowish-white or light yellowish-green, with no other longitudinal lines except for the light green dorsal lines, which are like dorsal blood vessels, and the spiracle is light dark brown. There are 28–38 crochets of abdominal legs arranged neatly, which are in an oval single-sequence ring. There are 21–31 crochets of tail legs, which are in an eyebrow-shaped semi-ring (Fig. 4-37).

Fig. 4-37 Larvae of *Scirpophaga incertulas* (Walker)

④ Pupa: The male pupa is about 12 mm long, thin, yellowish-white at first, and then yellowish-green. The head is small, and the antennae are about 7/8 of the wing length. The end of the forewing reaches the trailing margin of the fourth abdominal segment, the median leg extends beyond the end of the wing to the fifth abdominal segment, and the hindleg extends to the seventh or eighth abdominal segment. The female pupa is about 15 mm long, and its body is thick. The antennae are 1/2 of the length of the wings. The end of the forewing reaches the trailing margin of the fourth abdominal segment, the median leg is closer to the end of the wing, and the hindleg extends to the trailing margin of the sixth abdominal segment.

(2) Identification of damage features.

During the seedling stage and tillering stage of rice, the larvae bore into the base of rice stems, bite off the tissues in the stems, destroy the conduction function, which makes the inner leaves dehydrated and roll longitudinally like a shallot, and then gradually wither and yellow, resulting in withered seedlings. In the early stage of withered seedlings, the leaf sheath is not withered and yellow, does not twist and curl, and is easy to pull up, with flush insect bite marks at the fracture. The larvae invade at the end of the booting stage and the early heading stage, biting off the ear neck, which interrupts the nutrient transportation of rice, resulting in white ears. This kind of white ear is easy to be pulled up, the sword leaf sheath is not withered and yellow. When the leaf sheath is peeled off, there are flush and ring scars on the ear stem that are bitten off. When the ear stem is peeled off, there are a few insect scraps and feces in the stem, and the stem is green, white, and dry.

(3) Occurrence regularity.

Three generations occur in Jiangsu in a year, and incomplete fourth generations may occur in warm autumn. The mature larvae overwinter by diapause in rice piles, and begin to pupate when the temperature rises above 16°C from April to May of the following year. The first-generation imagoes are in full incidence in mid-May, peak in late May, and end of full incidence in late May and early June, causing damage to medium and late rice seedling fields and early planting fields, and resulting in withered inner leaves. The second-generation imagoes are in full incidence in early July, peak in mid-July, and end of full incidence in late July, damaging medium and late rice seedling fields and the seedlings of triple-cropping rice such as after-melon rice, and causing withered inner leaves. The third-generation imagoes are in full incidence in early August, peak in mid-to late August, and end of full incidence in late August and early September, causing white ears.

(4) Biological characteristics.

The eclosion of *Scirpophaga incertulas* (Walker) imagoes mostly occurs in the evening. They rest among the rice bushes during the day and begin to move at dusk. They have a strong phototaxis, most of which pounce on the light in the first half of the night, mostly on sultry nights with quiet wind and moonlight. They begin mating on the night of eclosion and lay eggs on the night of the next day. The oviposition of *Scirpophaga incertulas* (Walker) has obvious selectivity to the growth period of rice. The transplanting fields are more selected than the seedling beds, and the paddy field in the tillering stage and booting stage obviously attracts *Scirpophaga incertulas* (Walker) for oviposition. Under the same growth period, there are more eggs laid in densely grown fields with dark green leaves. The eggs of *Scirpophaga incertulas* (Walker) mostly hatch in the morning. The hatched borer crawls to the leaf tip, and then spins and hangs with the wind, scattering on the nearby rice plants. The larvae of *Scirpophaga incertulas* (Walker) are afraid of light, and the hatched borer first finds an appropriate part to bore into the rice plant and then feeds. The transplant of 2–3 instar larvae is also carried out at night, while the 3–4 instar larvae are mostly attached to a 2-cm-leaf (or stem) to crawl and transfer to other plants. The larvae only feed on the yellowish-white tender tissue of the leaf sheath, the pollen and stigma in the ear buds and the inner wall of the stem, and basically do not eat the parts with chlorophyll. Before feeding in large quantities, larvae will bite off most of the vascular bundles in a ring shape at the base between the leaf sheath and the rice stem node, resulting in withered inner leaves or white ears. The overwintering larvae in the rice stem move down to 1–2 cm below the soil before autumn harvest to diapause and overwinter, and pupate from April to May of the following spring.

Humidity has a great impact on the overwintering mortality rate of *Scirpophaga incertulas* (Walker). If the humidity is too low, the overwintering larvae in the rice piles on the soil surface of winter plowing fields are easy to dry to death, and if the humidity is too high, they are easy to die of suffocation and mildew. The predatory natural enemies of *Scirpophaga incertulas* (Walker) include frogs, dragonflies, spiders, etc. The parasitic enemies include a variety of parasitic wasps, parasitic flies, nematodes, etc. Among parasitic wasps, egg parasitic wasps are the most important. The main parasitic wasp species in Jiangsu include *Trichogramma japonicun* Ashmead, followed by *Telenomus dignus* Gahan and *Telenomus rowani* Gahan.

(V) Other rice pests

1. *Parnara guttata* Bremer et Grey

(1) Morphological identification.

The front surface of the wings is brown, and there are 7–8 translucent white spots arranged in a semi-circular shape on the forewings. There are 4 transparent spots in the center of the hindwings arranged in a straight line. The opposite side of the wing is light in color and covered with yellow powder. The stripes are similar to those on the front side of the wing. The two spots at the median cell end of the male butterfly are basically the same in size, while the upper one of the female butterfly becomes big and the lower one mostly degenerates into a small spot or disappears.

① Imago: It has a body length of 16–20 mm, and a wing span of 36–40 mm. The wings are dark brown with golden luster. There are 7–8 semi-circular white spots on the forewings and 4 white spots on the hindwings arranged in a "—" shape (Fig. 4-38).

Fig. 4-38 Imagoes of *Parnara guttata* Bremer et Grey

② Egg: The eggs are hemispherical, about 1 mm in diameter, flat at the top, slightly concave in the middle, with hexagonal indentations on the surface. They are light green at birth, then turn brown, and turn purplish-black when it is about to hatch.

③ Larva: The mature larva is 30–40 mm in length. The head is large, with W-shaped blackish-brown lines on the front. The first and second segments of the trunk are small, like the neck; the middle segment is hypertrophic, and the end is small, so the body is slightly spindle-shaped (Fig. 4-39).

(2) Identification of damage features.

Early damage causes white ears and yield reduction, and late damage causes

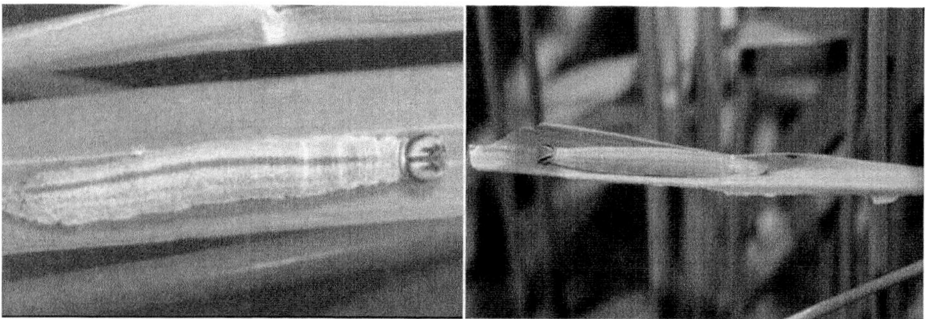

Fig. 4-39 Larvae of *Parnara guttata* Bremer et Grey

a large number of green leaves to be swallowed, resulting in a sharp decrease in green leaf area, insufficient paddy filling, low mass per thousand grains, and serious yield reduction. More seriously, due to the damage by *Parnara guttata* Bremer et Grey, the smut disease of rice grains aggravates sharply, and there are many diseased grains in the harvested rice. It is difficult to remove grains with smut during processing, which directly affects the quality of rice, causes economic losses, and threatens the health of consumers. In addition to rice, *Parnara guttata* Bremer et Grey also eats crops such as sugarcane, corn, wheat, and water bamboo, and can feed and survive on various weeds such as weeds, reeds, and barnyard. It damages the rice plants before heading, causing the rice ears to curl and unable to be drawn out, or twisted and unable to flower and bear fruit, seriously affecting the yield. In China, except in the northwest region, it is widespread in all rice areas, especially in the south of the Huaihe River basin. The hosts include rice, water bamboos, barnyards, weeds, reeds, etc. The larvae spin and bind several leaves of rice into buds, then hide inside and encroach on the leaves. In severe cases, they can eat up the leaves and often make the rice clusters unable to extend.

(3) Occurrence regularity.

In China, the main damage is caused by *Parnara guttata* Bremer et Grey, which occurs intermittently and seriously in some areas. Larvae in the southern rice area usually overwinter on *Leersia hexandra* Sw. and other weeds of Poaceae in wind-sheltered and sunny fields, ditches, ponds, lakes, shoals, low-humidity grasslands, etc., or overwinter in late rice clusters, root clusters at the lower part of ratoon rice, or water bamboo sheaths. Imagoes are active during the day, often suck honey on various flowers during the day, and eggs are scattered on rice leaves. Therefore, they occur in a large number in mountainous paddy fields, new rice fields, rice-cotton intercropping areas, or lakeside areas, causing serious damage. *Parnara guttata* Bremer et Grey occurs 6–8 generations a year in Guangdong, Hainan, and Guangxi; 5–6 generations a year in the south of Yangtze River basin and north of the Nanling area, such as Hubei, Jiangxi, Hunan, Sichuan, and Yunnan; 4–5 generations a year in the north of the Yangtze River basin; 3 generations a year in the north of the Yellow River basin; 2 generations a year in Liaoning. In single-cropping middle rice areas in Hunan, Jiangxi, Sichuan, Guizhou, and Hubei, *Parnara guttata* Bremer et Grey mainly causes damage from late June to July, especially causing damage to middle rice in mountainous areas. A large area of single-cropping late rice in lakeside areas is often damaged.

(4) Biological characteristics.

Imagoes are active during the day and like to suck nectar on plants such as sesame, pumpkin, cotton, and *Gomphrena globosa*, so the occurrence degree of the next-generation larvae can be predicted according to the number of imagoes on these plants. Eggs are scattered, mostly near the midrib on the back of leaves on rice; the number of eggs laid in paddy fields at tillering stage with thick green leaves and lush growth is large. The larvae have five instars. The first-instar and second-instar larvae bite an incision near the edge of the leaf tip, and then spin to roll the leaf margin into a bractlet. The number of leaves rolled by the third-instar larvae increases; generally 2–8 leaves are rolled into a bractlet. A larva can eat more than 10 rice leaves, and its food intake increases greatly after the fourth instar, accounting for more than 93% of food intake in its lifetime, so it should be controlled before the peak period of the third instar. The mature larvae pupate in the bract, and both ends of the pupae are tight and spindle-shaped. This pest is an intermittently rampant pest, and it mostly occurs with an optimum temperature of 24–30°C and relative humidity of more than 75%. In June and July, there is heavy rainfall and many rainy days, especially the "sunny and rainy alternatively, blowing southeast wind and daytime rain" can be used as a sign of major occurrence. High-temperature dry weather is not conducive to its occurrence. In mountainous areas or areas where rice is alternately planted with sesame, cotton, and other crops, the nectar source is abundant, thus *Parnara guttata* Bremer et Grey occurs seriously. In the egg stage, parasitic wasps play a great role, and the important ones are *Trichogramma japonicun* Ashmead, *Trichogramma confusum* Viggiani, etc.; the important natural enemies in the larval stage are *Apanteles ruficrus* (Haliday), *Itoplectis naranyae* (Ashmead), etc. Predatory natural enemies include a variety of spiders and ground beetles.

2. *Nephotettix bipunctatus* (Fabricius)

(1) Morphological identification.

The imago is 4.5–6 mm long. Length from head to wing end is 13–15 mm. There are many types of family members, and the biggest characteristic is that there are two rows of stiff spines on the tibia of the hindlegs. The imago is yellowish-green. The head is as wide as the pronotum, protruding forward in an obtuse round angle. There is a black transverse concave groove between the compound eyes on the vertex near the costal margins, with a black submarginal transverse band inside. The compound eyes are dark brown, and the ocelli are yellowish-green. The frontoclypeal area of the male is black, and the clypeus anterior and cheek area are light yellowish-green; the face of the female is light yellowish-brown, with several light brown transverse lines on both sides of the base of the frontoclypeus, and the cheek area is light yellowish-green. The pronotum of the female and the male is yellowish-green. The scutel is yellowish-green. The forewings are light bluish-green, the costal margin area is light yellowish-green, the wing end of the male is black at 1/3, and the female is light brown. The thorax, ventral surface, and back of the abdomen of males are black, while the ventral surface of females is light yellow and the back of the abdomen is black. Each leg is yellow. The eggs are eggplant-shaped and about 1–1.2 mm long; the last-instar nymphs are 3.5–4 mm long, and there are 4 instars in total (Fig. 4-40).

(2) Identification of damage features.

Nephotettix bipunctatus (Fabricius) stabs the stem and leaf of the hosts when feeding and ovipositing, destroying the conducting tissues with brown stripes on the damaged parts, resulting in yellow or withered plants.

Fig. 4-40 Imagoes of *Nephotettix bipunctatus* (Fabricius)

(3) Occurrence regularity.

Five to six generations occur every year in Jiangsu and Zhejiang, and 3rd–4th instar nymphs and a few imagoes overwinter on weeds by green manure fields, ponds, and rivers. Imagoes lay eggs in the inner tissue of the leaf sheath margin, with more than 100–300 eggs per female. Nymphs like to inhabit the lower part of the plant or the back of the leaf to feed, which are clustered. The 3rd–4th instar nymphs are especially active. The overwintering nymphs mostly become imagoes by eclosion in April and migrate to paddy fields or water bamboo fields for damage. They are easy to occur in rainless years. The main natural enemies are *Paracentrobia andoi* (Ishii) and predatory spiders.

3. *Echinocnemus squameus* Billberg

(1) Morphological identification.

① Imago: Imagoes are 5 mm long, dark brown, and densely covered with grayish-brown scales. The head is elongated like an elephant trunk, and the antennae are blackish-brown, with enlarged ends, and are attached to the elephant-trunk snout at the proximal end. There are 10 longitudinal grooves on each of the two wing sheaths, and there is a long small white spot below each (Fig. 4-41).

② Egg: The egg is oval, 0.6–0.9 mm long, milky white at birth, and then light yellow, translucent, and glossy.

③ Larva: The larva is 9 mm long, maggot-shaped, slightly curved to the ventral surface, fat and wrinkled, brown head, milky white thorax and abdomen, much like white rice.

Fig. 4-41 Imagoes of *Echinocnemus squameus* Billberg

④ Pupa: The pupa is about 5 mm long, milky white, then gray, with fine wrinkles on the ventral surface.

(2) Identification of damage features (Fig. 4-42).

The imagoes bite the stems and leaves of seedlings with a tubular snout. After the damaged inner leaves are taken out, the lightly damaged leaves of the seedlings present a row of small holes, and the heavily damaged leaves are broken and float on the water's surface. The larvae feed on the tender fiber roots of rice plants, resulting in the yellowing of leaf tips and poor growth. In severe cases, the heading is not allowed, blighted grains are caused, or even rice plants are withered in pieces.

Fig. 4-42 Damage Features Caused by *Echinocnemus squameus* Billberg

(3) Occurrence regularity.

One generation occurs annually in Zhejiang, one generation occurs annually in some areas of Jiangxi and Guizhou, while two generations occur annually in Guangdong. *Echinocnemus squameus* Billberg in first generation area overwinters as imagoes, and the one in the intersection area of the first and second generations and the second generation area overwinters mostly as imagoes. Larvae can also overwinter, and some overwinter as pupae. The larvae and pupae mostly overwinter in the rhizosphere 3–6 cm from the soil surface, and the imagoes often hibernate and overwinter under the weeds and fallen leaves on field ridges and ground edges. The overwintering imagoes in southern Jiangsu lay eggs from May to June of the following year and eclosion often occurs in October. The overwintering imagoes in Jiangxi lay eggs in early and mid-May and incubate the first-generation larvae in late May, and eclosion occurs from mid-July to mid- and late August. The second-generation larvae hatch from late July to early and mid-August, and some pupate or overwinter after eclosion in October. Generally, the damage is the most intense during the early rice regreening period.

(4) Biological characteristics.

Imagoes are active in the morning and evening and hide in the seedling field or at the base of the rice bushes, or in the grass at the field ridge during the day, with feign death and phototaxis. Before eggs are laid, a small hole is bitten on the rice stem or leaf sheath about 3 cm above the water surface, and 13–20 eggs are laid in each hole. After hatching, the larvae stay in the leaf sheath for a short time to feed, and then bore into the soil along the rice stem, generally clustering at a depth of about 2–3 cm under the soil to feed on the tender fiber roots and humus of rice. A clump of rice roots mostly has dozens of larvae damaging the roots. Generally, the situation mostly occurs in hills, semi-mountainous areas than plains, dry fields in high altitudes than low-lying fields, and sandy soil than clayey soil. The larvae prefer to gather under the soil and feed on tender rice roots. After maturity, a soil chamber is built 3–7 cm under the soil near the rice roots for pupation. Sandy loam fields, dry fields, and dry seedling fields with good aeration and low water content are vulnerable to damage. Warm and rainy weather in spring is conducive to pupation and eclosion, and rainy conditions in the tillering stage of early rice are conducive to egg laying by imagoes. One to two generations occur annually, generally one generation in the single-cropping rice area, and two generations in the double-cropping rice area or the mixed planting area of single-cropping rice and double-cropping rice. *Echinocnemus squameus* Billberg mainly overwinters around rice piles and in soil gaps as imago, but some also overwinter in ridges, furrows, grass, and loose soil. A few overwinter in soil chambers 3–6 cm below the soil near rice piles as larvae pupation.

4. *Naranga aenescens* Moore

(1) Morphological identification.

① Imago: The imago body is dark yellow. The male moth has a body length of 6–8 mm and a wing span of 16–18 mm. The forewings are dark yellowish-brown, with two parallel dark purple wide oblique bands. The hindwings are grayish-black. The female moth is slightly larger and its body color is lighter than the male moth. The forewings are light yellowish-brown, and the middle of the two purplish-brown oblique bands is disconnected; the hindwings are grayish-white (Fig. 4-43).

②Egg: The eggs are oblate, with vertical and horizontal carinas on the surface, forming many trellis designs. They are light yellow at birth and turn purple before hatching.

③Larva: The larva is about 22 mm, green, the head is yellowish-green or light brown, the dorsal line and subdorsal line are white, and the spiracle line is yellow. There are only two pairs of pleopoda and a pair of caudal legs, and the larva looks like an inchworm when walking (Fig. 4-44).

④Pupa: The pupa obtecta is initially green and gradually turns yellowish-brown. There are 4 pairs of crochets at the end of the abdomen, and the last pair is the longest.

(2) Identification of damage features.

In addition to rice, *Naranga aenescens* Moore also damages sorghum, corn, sugarcane, and water bamboo and feeds on a variety of weeds of Poaceae. The larvae feed on rice leaves. The larvae of 1st–2nd instars feed on leaves into white stripes. After the 3rd instar, they feed on leaves into incisions. In severe cases, they bite the

Fig. 4-43 Imagoes of *Naranga aenescens* Moore

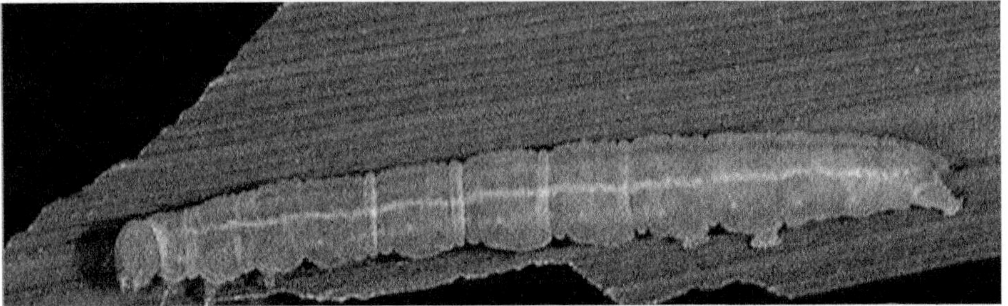

Fig. 4-44 Larva of *Naranga aenescens* Moore

leaves into pieces, leaving only the midribs. The seedling stage is the most damaged.

(3) Occurrence regularity.

Naranga aenescens Moore occurs 6–7 generations in Guangdong in a year, and it overwinters as a pupa in rice stubbles or leaf wraps and sheaths of rice sticks and weeds. In July and August, it mostly damages the late rice seedling fields, and the population density is generally low in other months. It occasionally damages the early rice in the tillering stage in April and May.

(4) Biological characteristics.

Imagoes lurk in rice stems and leaves or grasses during the day, copulate and oviposit at night, with strong phototaxis, and most of them are female moths that do not oviposit under the lamp. Eggs are mostly laid in the middle of rice leaves, and a few eggs are laid in leaf sheaths. Each egg mass generally has 3–5 eggs arranged in 1 or 2 rows, and some eggs are laid individually, with an average of about 500 eggs per female. The leaves of rice seedlings are dark green, which can attract imagoes to oviposit in a concentrated manner.

The larvae start to feed about 20 minutes after hatching, feed on the leaf tissue first, and gradually gnaw the chlorophyll, resulting in yellow linear spots on the leaf surface. After the 3rd instar, they bite from the leaf margin and bite the leaves into incisions.

When the larvae move on the leaves, they jump into the water when they are disturbed, and then swim or climb to other rice plants to cause damage. The older the instar, the greater the food intake, eventually leaving only one midrib on the leaves. The mature larvae spin silk at the tip of the leaf to twist the rice leaves into pyramid-shaped triangular bracts, hide in the bracts, bite off the leaves, make the bracts float on the water surface, and then cocoon and pupate in the bracts.

Section II Green Prevention and Control Technology for Integrated Farming in Paddy Fields

In the farmland ecosystem of the integrated farming mode of machine-transplanted paddy fields: First, due to the addition of active animals such as ducks, fish and crayfish, the ecological food chain in the farmland is extended, which is conducive to the survival and reproduction of natural enemies and other populations, making a great change in the original single machine-transplanted rice farmland system and making it more biodiversity; Second, the activities of ducks, fish, and crayfish in the field continuously stimulate the epidermis of rice plants in the field, which changes the invasion, spawning, and hatching of pests, as well as the infection, reproduction, and transmission of pathogenic bacteria of diseases; Third, the residues, residual liquid, and excrement generated after feeding ducks, fish, and crayfish make the water in the farmland more enriched and nutritive than that in the single machine-transplanted rice fields, and the community of plankton and beneficial bacteria increases; Fourth, the increased plant and row spacing and reduced density by the integrated farming mode of machine-transplanted paddy fields are conducive to aeration and light transmission at the base of rice, enhancing the photosynthesis of individual plants, and facilitating the robust growth of rice; Fifth, the water layer is maintained in the field for a long time, so it cannot drain the water in fields, and the rice base is soaked in water for a long time, which is not conducive to the aeration of the rice root system; Sixth, rice-duck, rice-fish, rice-crayfish, etc. coexist in water, and in such cases, chemical pesticides cannot or are not easy to be used before and in the middle of transplanting in the field to enhance safety, which makes the prevention and control of diseases and pests more difficult.

I. Understanding of the Concept of Green Prevention and Control for Integrated Farming in Paddy Fields

Starting from the overall rice farmland ecosystem and according to the interrelationship between pests and the environment, give full play to the role of natural control factors, coordinate the application of necessary measures according to local conditions, and take various technical measures on the basis of the concept of "public plant protection and green plant protection" and in accordance with the plant protection policy of "prevention first and comprehensive control" to reduce pathogenic bacteria, the base of pest sources, hazards, and the probability of outbreaks, and control pests below the allowable level of economic damage, so as to obtain the best economic, ecological, and social benefits.

II. Technical Strategies for Green Prevention and Control for Integrated Farming in Paddy Fields

Take plant quarantine as the premise, agricultural control measures as the basis, and seed treatment as the key to reducing the source of diseases and the base of pest sources. Take the protection of the farmland ecosystem in the field as the core, supplemented by ecological regulation, biological control, and scientific medication to reduce the occurrence of diseases and pests and the probability of outbreaks, and ensure the safety of integrated rice-duck, rice-fish, and rice-crayfish farming in paddy fields.

III. Technical Measures for Green Prevention and Control for Integrated Farming in Paddy Fields

1. Green prevention and control technology before rice sowing

(1) Strict implementation of the plant quarantine system.

In the process of introduction and transportation of varieties, quarantine should be strengthened to control the introduction, occurrence, transmission and spread of quarantine diseases and pests such as *Xanthomonas oryzae* pv.*oryzicola* (Fang et al.) Swings and *Xanthomonas oryzae* pv.*oryzae* (Ishiyama) Swings from the source.

(2) Selection of high-quality and high-yield disease-resistant varieties.

The resistance of varieties plays a decisive role in the occurrence and development of diseases and pests in the field. Special attention should be paid to rice varieties selected for integrated farming in paddy fields having high or medium resistance to major diseases and pests such as rice blast, sheath blight, *Ustilaginoidea virens* (Cooke) Takahashi and rice planthopper, so as to reduce the pressure of chemical control in seedling fields and transplanting fields.

2. Green prevention and control technology in the rice seedling stage

Gibberella fujikuroi (Saw.) Wollenw and *Aphelenchoides besseyi* Christie are two important seed-borne diseases on rice. They can only be prevented because once they occur, they cannot be controlled by pesticides. Therefore, seed treatment must be done well before sowing.

(1) Seed treatment.

① Sunning seeds before sowing.

The selected rice seeds are spread on a cement field in sunny weather for 1–2 days of exposure 2–3 days before sowing. First, ultraviolet rays can kill pathogens and eggs transmitted by seed with a bacterium, such as rice blast pathogens. Second, seed activity can be enhanced, and seed germination and emergence rate can be improved.

② Chemical-seed mixing.

The new seed treatment technology is based on the achievements of "integration and promotion of whole-process simplified prevention and control technology for rice diseases and pests". Specific steps: First, soak the rice seeds according to the routine. Second, drain them, and then directly mix them with chemicals. 4 kg seeds are used per mu with the preparation of 30 mL of 20% chlorantraniliprole, 100 mL of 24% thifluzamide, and 20 g of 10% triflumezopyrim. It should be noted that no water is required during chemical-seed mixing. After mixing, red ink can be used as the indicator color, and the seeds can be sown after drying.

This technological achievement won the first prize for agricultural technology promotion in Jiangsu Province in 2020. For the chemical-seed mixing of this formula, the chemicals have a long effective period, which can effectively control seed-borne diseases such as bakanae disease and *Aphelenchoides besseyi* Christie, and can effectively control the occurrence of diseases and pests at the early and middle stages after machine transplanting of rice, such as rice planthoppers, *Cnaphalocrocis medinalis* (Guenée), borers, leaf (seedling) blast, rice stripe virus, and sheath blight. According to many years of experiments by the Zhenjiang Academy of Agricultural Sciences, the effect of pest control and disease suppression can reach 70–90 days, which reduces the number of applications at the early and middle stages of transplanting in the field and protects the ecological environment of the farmland.

(2) Cultivation of strong disease-free and pest-free seedlings.

Industrialized seedling cultivation should be adopted, and attention should be paid to adjusting the sowing and transplanting time to avoid the peak migration of *Laodelphax striatellus* (Fallen) and the peak stage of oviposition of borer. The greenhouse of the seedling cultivation should be blocked by insect-proof screens to block contact between imagoes such as *Laodelphax striatellus* (Fallen) and borers and host seedlings, so as to control the occurrence of rice stripe virus and damage to seedlings.

3. Green prevention and control technology in the rice field period

On the basis of agricultural control measures, the long-acting characteristics of seed treatment agents are applied to effectively control or reduce the occurrence of diseases and pests such as bakanae disease, *Aphelenchoides besseyi* Christie, and *Laodelphax striatellus* (Fallen) at the early and middle growth stages of rice, so as to protect the ecological environment for the reproduction and survival of natural enemies such as frogs, spiders, and *Apanteles* spp. in the field, improve the source and breeding environment of diseases and pests, and enhance natural damage control ability and crop resistance to diseases and pests. At the same time, physical and chemical induction and control techniques such as planting phanerogam, vetiver, water bamboo, sex attractants, as well as light trapping and killing are used to trap imagoes such as field borers and *Cnaphalocrocis medinalis* (Guenée) to mate and lay eggs, so as to reduce the occurrence base of eggs and the number of pesticide applications, protect the ecological environment of farmland, and achieve the effect of natural damage control by natural enemies in farmland. The application of green prevention and control biological pesticides, such as matrine, *Empedobacter brevis*, and validamycin, is promoted to prevent and control *Cnaphalocrocis medinalis* (Guenée) and sheath blight, and high-efficiency, low-toxicity, and low-residue pesticides are selected to prevent, and control major diseases and pests such as rice planthopper and rice blast in emergent years.

(1) Agricultural control technology.

① The fields are selected far away from old infected areas and seriously diseased fields, so the occurrence will be controlled from the source.

Overall planning, reasonable arrangement of croppings, and contiguous planting should be conducted, and mixed planting of rice with different cropping systems should be avoided as far as possible. By doing so, epidemic diseases and intermediate hosts of migratory pests and the occurrence degree will be reduced.

② The fields should be cleaned, so as to eliminate the source of overwintering diseases and pests and reduce the base of disease and pest sources.

It is necessary to remove weeds and residual plants in fields and ditches, pick up and destroy rice piles in the field, destroy straw, and reduce the overwintering base of *Laodelphax striatellus* (Fallen), borer, and rice blast pathogen sources.

③ Winter plowing and sunning of the upturned soil, spring plowing, and irrigation should be carried out to destroy the overwintering places with endogenous diseases and pests.

In winter leisure, if green manure planting fields need to be cultivated, winter plowing and sunning of the upturned soil should be carried out as early as possible to destroy the overwintering places of diseases and pests, especially the borer. Before pupation in spring, plowing and irrigating deep water fields should be carried out in time to kill the pupae in rice roots and reduce the source of pests.

④ When the transplanting field is leveled, sclerotium and weed seeds at the edge of the field should be salvaged and cleaned.

In addition to plowing the deep-buried sclerotium and weed seeds, when paddy fields are leveled, sclerotium and weed seeds should be salvaged and cleaned in time at the corner of the field at the lower air port of the field, so as to reduce the occurrence base of *Rhizoctonia solani* Kühn, *Ustilaginoidea virens* (Cooke) Takahashi, etc. and weed seeds.

⑤ Fertilizer and water management should be strengthened to promote the robust growth of rice seedlings and enhance their stress resistance.

Scientific and reasonable fertilization should be achieved. Based on testing soil for formulated fertilization, organic fertilizer should be added to the basal fertilizer, and N, P, and K should be reasonably combined for fertilization. If conditions permit, the technology of machine transplanting, together with the side deep fertilization and frequent shallow water irrigation, should be carried out to create a suitable environment for the robust growth of rice seedlings and continuously enhance the disease resistance, pest resistance, and fall resistance of rice itself.

(2) Ecological regulation technology.

① Control the use of pesticides at the early stage to protect the ecological and natural circulation balance of the farmland.

Make full use of natural ecological communities such as rice-duck, rice-fish, and rice-crayfish to maintain biodiversity and protect the ecological and natural circulation balance of farmland. Protect and utilize the self-control effects of natural enemies such as frogs, spiders, lacewings, ladybugs, assassin bugs (Reduviidae), parasitic wasps, and parasitic flies.

② Plant phanerogam and build "bridge fields" to transfer natural enemies.

Phanerogams such as sesame, soybeans, peas, and flowers are planted on the ridges and roadsides of the fields to build a "bridge field" for the transfer of natural enemies in the field. Food and honey sources are provided to allow natural enemies to multiply and reproduce, so as to provide a "supplementary" source for the source of natural enemies in the early and middle stages of the next transplanting.

(3) Physical and chemical induction and control technology.

① Plant vetiver and water bamboo to lure.

By taking advantage of the green-taxis of pests, tall and lush plants such as vetiver (Fig. 4-45) and water bamboo are planted in the water sources of existing ditches, canals, ponds, and dams to replace the hosts, so as to lure the imagoes of pests such as borer, *Cnaphalocrocis medinalis* (Guenée) and rice planthopper to intensively lay eggs, incubate and damage, thus reducing the occurrence base in the field.

Fig. 4-45 Vetiver

② Application of sex attractant trapping.

By artificially synthesizing some sex pheromone chemicals released by female moths after sexual maturity, male moths of the same species seeking to mate in the field are attracted and killed in the traps, so that the females lose the opportunity to mate and cannot reproduce their offspring effectively, the population of offspring is thus reduced, then the purpose of pest control is reached. In recent years, this technology has been widely used in integrated farming of paddy fields. Generally, there are 20–30 traps installed along the field ridge per mu, which can effectively trap and kill the imago of various pests such as rice borers and *Cnaphalocrocis medinalis* (Guenée) and reduce the occurrence base in the field.

③ Light trapping.

Black light lamps are most commonly used to kill imagoes through their phototaxis. Insecticidal lamps, mainly aimed at the imagoes, are effective in the early stage of rice planting but are weak in the face of pests that have reproduced to a large scale. In addition, insecticidal lamps have a poor selection of insect species, killing pests as well as many beneficial insects and natural enemies, which remains a great technical controversy on its technical feasibility. Therefore, in practical use, it is necessary to select the light source spectrum that is targeted to the pests, and strictly control the light-on time, so as to trap and kill pests such as rice planthoppers, *Cnaphalocrocis medinalis* (Guenée), and borers, and protect their natural enemies.

(4) Biological control technology.

① Artificial release of natural enemies.

In contiguous planting areas where fields are relatively concentrated and the condition permits, predatory natural enemies such as *Trichogramma* can be artificially released to supplement the species and quantity of natural enemies in the farmland, thus achieving the effect of natural ecological pest control.

② Application of microbial pesticides.

Fungi, such as *Metarhizium anisopliae* and *Beauveria bassiana*, are the most widely used and mature biological pesticides in integrated farming of paddy fields at present, and have good control effects on pests such as borers, *Cnaphalocrocis medinalis* (Guenée), and thrips.

Bacteria, such as *Bacillus thuringiensis* (Bt), *Empedobacter brevis*, *Bacillus cereus,* and *Bacillus subtilis*, have a certain control effect on *Cnaphalocrocis medinalis* (Guenée), borers, sheath blight, etc.

Viruses, such as nuclear polyhedrosis viruses (NPV), have a certain control effect on lepidopteran pests.

③ Selection of botanical pesticides.

Botanical pesticides are compounds and derivatives derived from active ingredients extracted from plants and synthesized according to the plant itself and its active structure. For example, matrine, osthole, azadirachtin, pyrethrum, santonin, lentinan, triptolide, etc. can effectively control various plant diseases and insect pests, such as lepidopteran pests in paddy fields.

④ Application of biochemical pesticides.

Such pesticides are mainly antibiotics, including validamycin, kasugamycin, ningnanmycin, Antimycoin 120, etc., which have good control effects on rice sheath blight, rice blast, *Ustilaginoidea virens* (Cooke) Takahashi, rice stripe virus, etc.

It should be noted that biological pesticides themselves do not have adsorbability and are mostly contact-killing, so they should be evenly sprayed when used. Most biological pesticides are active proteins or cells, which have strict requirements for temperature, humidity, and other conditions, and cannot be mixed with concentrated alkaline drugs. Since biological pesticides, whose action mechanism differs from that of chemical pesticides, require parasitism and reproduction to function, they are applied 3–5 days before the pest formation. In most cases, biological pesticides are applied in the peak stage of egg hatching to ensure a control effect.

(5) Emergency prevention and control technology.

In the middle and late stages of rice, in the case of years when epidemic diseases such as rice blast and rice false smut, as well as migratory pests such as *Cnaphalocrocis medinalis* (Guenée) and *Nilaparvata lugens* (Stål) occur repeatedly or greatly, emergency prevention and control should be carried out under the technical guidance of local plant protection departments, and aimless application of drugs should be prevented to avoid losses.

First, the continuous investigation of diseases, pests, and natural enemies in the field should be carried out, and pesticides should not be applied until the control indicators are met according to the economic threshold evaluation.

Second, pesticides must not be applied before live aquaculture, such as ducks, fish, and crayfish, are harvested or separated from rice areas with a strict isolation zone to ensure their safety.

Third, select safe, efficient, low-toxicity, and low-residue pesticides. Biological pesticides and their complex are selected as far as possible, which are not only safe and effective but also avoid pesticide residues and prevent the killing of natural enemies.

Fourth, select the correct application method and time, so as to coordinate the application method with the action mechanism and pesticide formulation, with the parts damaged by diseases and pests, and with the protection and utilization of natural enemies.

Fifth, the drift of liquid pesticides in the process of application and the cross-flow of water sources in the applied area and isolation area are avoided to ensure the safety and effectiveness of pesticide application.

Chapter V Yield and Benefit Analysis of Integrated Farming Mode in Paddy Fields

In recent years, with the rapid development of the social economy and the continuous improvement of people's quality of life in China, people have increasingly strict requirements for food safety and quality, and the demand for high-quality rice and aquaculture products is also growing, resulting in more requirements for agricultural development. Integrated farming in paddy fields conforms to social development demands. Such a symbiotic and complementary ecological agricultural mode of planting and breeding will promote breeding by planting, supplement planting with breeding, realize comprehensive utilization and circular development, which changes the traditional single planting or breeding mode in the past, improves the land utilization rate, and greatly improves the ecological environment of farmland. Integrated farming in paddy fields cuts off the effective transmission of harmful microorganisms, which not only reduces the prevalence of diseases and weeds in the farming of plants and animals but also reduces the usage of chemical fertilizers and pesticides to effectively control rural non-point source pollution. Meanwhile, it saves energy and resources, effectively promotes the development of ecological circular agriculture, and realizes the good effects of multiple uses of one field, multiple uses of one water source, and multiple harvests in one season.

Under the current situation of vigorously advocating green development, changing the mode of agricultural growth, and advancing the structural reform on the agricultural supply side, various regions all over China, based on this green, high-quality, and efficient mode of rational resource utilization and innovative planting and breeding, make full use of the advantages of paddy field resources to develop integrated farming in paddy fields, which not only ensures food security but also increases the income from rice planting.

The common development of economic, social, and ecological benefits has been realized.

Section I Economic Benefit

Affected by the scale of social planting and breeding and market consumption, the economic benefits of integrated farming in paddy fields may vary in different modes of planting and breeding, in different years of the same mode, and in sales modes where products are sold as primary or secondary products.

[According to the data from Taizhou City in 2020]

The total area of integrated rice-crayfish farming was 1,427 hm^2, of which 1,373 hm^2 was in Xinghua, accounting for more than 95%, and a small amount of which was distributed in Jiangyan, Jingjiang, and other places. Generally, paddy field yielded green rice 300 kg/mu, bringing an income of CNY 3,000 based on a unit price of CNY 10/kg; the same paddy field yielded crayfish 100 kg, bringing an income of CNY 3,800 based on a unit price of CNY 36–40/kg; the materialized cost was CNY 2,200, and the labor cost and land rent were CNY 1,800; thus the net benefit per mu was CNY 2,800, which was 4.6 times that of the conventional rice-wheat planting mode.

The total area of integrated rice-duck farming in the city was 793 hm^2, mainly distributed in Xinghua (586 hm^2), Jiangyan (146 hm^2), Taixing, Jingjiang, and other places (60 hm^2). Generally, paddy field yielded green rice 425 kg/mu. If calculated at CNY 10/kg, the income benefit per mu was CNY 1,850, which was 3 times the income of the conventional rice-wheat planting mode.

The total area of integrated rice-crab farming in the city was 186 hm^2, mainly distributed in Xinghua (100 hm^2) and Taixing (86 hm^2). Generally, paddy field yielded high-quality rice 250 kg/mu, bringing an income of CNY 5,000 based on a unit price of CNY 20/kg; the same paddy field yielded river crabs 50 kg, bringing an income of CNY 3,000 based on a unit price of CNY 60/kg; the materialized cost was CNY 2,450, and the labor cost and land rent were CNY 2,500; thus the net benefit per mu was CNY 3,050, which was 5 times that of the conventional rice-wheat planting mode.

The total area of integrated rice-fish farming in the city was 146 hm^2, including 60 hm^2 in Jiangyan, 26 hm^2 in Xinghua, 6 hm^2 in Taixing, and 53 hm^2 in Jingjiang. From the analysis of input-output benefits, generally, paddy fields yielded green rice 300 kg/mu, bringing an income of CNY 3,000 based on a unit price of CNY 10/kg, and the income of fish per mu was CNY 3,000; the materialized cost was CNY 1,850, and the labor cost and land rent were CNY 1,500; the net benefit per mu was CNY 2,650, which was 4.4 times that of the conventional rice-wheat planting mode.

[According to the data of Yizheng City in 2019]

Mode and scale of planting and breeding: In 2017, the area of integrated farming in paddy fields in Yizheng City was about 59.3 hm^2. The main 3 modes were rice-duck, rice crayfish, and rice-loach. The areas were mainly concentrated in villages and towns such as Chenji, Yuetang, Zaolinwan, Maji, Xinji, and Zhenzhou. Among those areas, the rice-duck scale was 46.0 hm^2, the rice-crayfish scale was 10.0 hm^2, and the rice-loach scale was about 3.3 hm^2. In 2018, the area of integrated farming in paddy fields in Yizheng City was about 159.3 hm^2, involving the whole city. The main 4 modes were rice-duck, rice-crayfish, rice-frog, and rice-fish, of which the rice-duck scale was 104.0 hm^2, rice-crayfish about 45.3 hm^2, rice-frog about 6.7 hm^2, and rice-fish about 3.3 hm^2 (Table 5-1).

Costs and benefits: In 2017, through integrated farming in paddy fields, the total yield of rice was 391.20 t, the total yield of other planting and breeding was 32.58 t, the total output value of rice was CNY 1,956,000, the total output value of other planting and breeding was CNY 1,082,400, the total investment was CNY 1,757,550, and the total income was CNY 1,280,850. In 2018, the total yield of rice was 1,065.6 t, the total yield of other planting and breeding was 97.52 t, the total output value of rice was CNY 6,393,600, the total output value of other planting and breeding was CNY 3,401,600, the total investment was CNY 5,501,560, and the total income was CNY 4,293,640.

Table 5-1　Costs and Benefits of Two Main Integrated Farming Modes of Paddy Fields in Yizheng City

Mode	Year	Area (hm^2)	Rice yield (kg/mu)	Rice price (CNY/kg)	Other yield (kg/mu)	Other prices (CNY/kg)	Cost (CNY/mu)	Output value (CNY/mu)	Net income (CNY/mu)
Integrated rice-crayfish farming	2017	10.0	400	5	70	40	2,425	4,800	2,375
	2018	45.3	420	6	70	40	2,876	5,320	2,444
Integrated rice-duck farming	2017	46.0	480	5	32	30	2,020	3,360	1,340
	2018	104.0	500	6	32	30	2,273	3,960	1,687

If taking single rice planting as the control, the yield of the control rice was 600 kg/mu, the output value was CNY 1,800/mu based on an average selling price of CNY 3/kg, and the net income was CNY 950/mu after deducting the production cost of CNY 850/mu. Through income analysis, it can be seen that paddy fields of integrated farming have a net increase of output value of more than CNY 150/mu compared with that of single rice planting. The net increase in income from integrated rice-duck farming is slightly lower, while that from integrated rice-crayfish farming is more than CNY 1,400/mu, with significant efficiency.

As there are numerous integrated farming modes, the following field investigation was conducted on the integrated rice-crayfish, rice-duck, and rice-*Trionyx* farming modes with relatively large farming areas and high benefits. The case analysis is as follows.

I. Integrated Rice-Crayfish Farming Mode

[Case 1] Chuanjie Family Farm in Jurong City (crayfish as the primary product)

The family farm was established in 2015 and was rated as a director unit of the Jurong Fisheries Association in 2020. Wu Chuanjie, as the legal representative of the farm, has been engaged in crayfish breeding and rice planting for 9 years. The farm has 2 farming bases with a total area of 14 hm^2 (8.67 hm^2 for one base in Taiping Village, Baitu Town, Jurong City, and 5.33 hm^2 for the other base in Wugang Village, Gaozi Town, and Zhenjiang High-tech Zone).

The planting and breeding benefits of this farm are affected by the scale of social planting and breeding and market consumption, which vary in different years. According to the investigation on the planting and breeding benefits of Chuanjie Family Farm in Jurong City in recent years, the income ranges from CNY 38,400–47,724/hm^2, with the best in 2020. Here, the planting and breeding situation of the base in Taiping Village, Baitu Town of Chuanjie Family Farm in Jurong City in 2020 is taken as an example for analysis.

(1) Process of planting and breeding.

From March 15 to March 27, fingerling crayfish were put in for the first time, with an input amount of about 315 kg/hm^2 (about 108,000 crayfish seedlings). From April 28 to June 20, crayfish were caught successively, with a catch amount of about 1,560 kg/hm^2. On June 22, rice seeds (Variety: Nanjing 9108) were sown in the paddy fields, with a sowing amount of about 90 kg/hm^2. On July 1, fingerling crayfish were put in for the second time, with an input amount of about 345 kg/hm^2. From August 1 to September 26, crayfish were caught successively, with a catch amount of about 570 kg/hm^2.

(2) Sales value.

From April 28 to June 20, a total of 13,520 kg of crayfish were caught in the base, which resulted in an output value of CNY 602,992 based on an average unit price of CNY 44.6/kg. From August 1 to September 26, a total of 4,940 kg of crayfish were caught in the base, which resulted in an output value of CNY 243,048 based on an average unit price of CNY 49.2/kg. In November, the rice was harvested, and the rice yield per unit area was 7,808 kg/hm^2, and the total rice yield was 62,460 kg (actual sown area of 8 hm^2 except circular ditches), which resulted in an output value of CNY 151,153 based on an average unit price of CNY 2.42/kg. The total output value of crayfish and rice was CNY 997,193.

(3) Cost input.

From March 15 to March 27, a total of 2,730 kg of fingerling crayfish were put in the base. Based on an average unit price of CNY 60/kg, the subtotal was CNY 163,800. On July 1, 2,993.5 kg of fingerling crayfish were put in for the second time, with a subtotal cost of CNY 47,896 based on a unit price of CNY 16/kg. The total input of feed was 11 t (Cargill 35 protein feed), with a unit price of CNY 6,500/t. The subtotal cost was CNY 71,500. The subtotal of drugs, animal health products, and electricity charges was CNY 23,000. The subtotal cost of rice seeds 780 kg was CNY 5,928 based on a unit price of CNY 7.6/kg. The subtotal of pesticide, chemical fertilizer, and UAV operation costs was CNY 31,000. The subtotal of the harvester operation cost was CNY 7,800. The subtotal of the land transfer fee paid was CNY 110,500 based on a rent of CNY 12,750/hm^2. Wages for permanent workers (2 workers) were CNY 80,000, and wages for temporary workers were CNY 42,000. The total input value was CNY 583,424.

(4) Economic benefits.

After subtracting the cost input from the planting and breeding output value of the base, the total income was CNY 413,769 (including the wage of the farm's legal person), and the income per hectare was CNY 47,724.

[Case 2] Jiangsu Shenhe Agricultural Development Co., Ltd. (rice and crayfish as the primary products)

In 2020, the company's area of integrated farming in paddy fields was 13.67 hm^2, and it adopted the integrated rice-crayfish farming mode.

(1) Farmland works.

The opening width of the circular ditch was 2.5 m, the ditch depth was about 1–1.2 m, and the water surface accounted for 10% of the paddy field area. Breeding density: 300 kg/hm^2 with a specification of 160 fingerling crayfish/kg. Breeding management: Crayfish were fed once in the morning and once in the evening, and the feeding amount accounted for 3%–8% of the crayfish weight, which increased day by day.

(2) Sales value.

In 2020, the output value of commercial crayfish was CNY 820,000 based on a yield of 1,500 kg/ hm^2 and an average unit price of CNY 40/kg; the output value of rice was CNY 806,000 based on a yield of 5,360 kg/hm^2 and a unit price of CNY 11/kg. The total output value of crayfish and rice was CNY 1.626 million.

(3) Cost input.

The total cost input for production, including labor, fingerling crayfish, feed, seeds, facilities, equipment, etc., was CNY 645,000.

(4) Economic benefits.

After subtracting the cost input from the planting and breeding output value of the base, the total income was CNY 981,000, and the income per hectare was CNY 71,727, which was 5.3 times that of conventional planting.

II. Integrated Rice-Duck Farming Mode

[Case] Houbai Farm in Jurong City (rice and adult duck as the primary products)

In 2020, the farm's area of integrated farming in paddy fields was 36.7 hm^2, and it adopted the integrated rice-duck farming mode.

(1) Farmland works.

The area of integrated rice-duck farming was located in contiguous paddy fields with fertile soil, convenient drainage and irrigation, and easy water conservation. In addition, the base had relatively complete facilities, so these fields could meet the requirements of high-standard farmland. Every 0.4 hm^2 of the field was enclosed into a breeding unit with a net, and the mesh size of the net was no more than 2 cm. Each unit was built with a duck shed of 3 m^2. Piles were nailed to buckle the net every 4 m or so, and the net edge was compacted on the base, and the net height was about 80 cm. Breeding density: In order to increase the economic benefits of duck breeding, local shelduck with a larger size were selected. After the transplanting and establishment of the rice seedlings (7–10 days), the ducklings aged 10–15 days were put into the field after water training, and 15 ducklings were bred per mu. The breeding quantity of each breeding unit was calculated by area. Breeding management: Young ducks were supplied with adequate mixed feed in the first 3 days after they were put into the fields (before that, the feed was put in the duck shed and made the ducks remember where they would be fed before going down to the fields), and then the feeding times were gradually reduced to 2–3 times a day; after 10 days, they were fed with wheat and other unprocessed grains, and after 20 days, they were fed once a day.

(2) Sales value.

In 2020, 7,600 adult ducks were sold at an average price of CNY 85/duck, bringing an output value of CNY 646,000; the paddy fields yielded rice 5,550 kg/hm^2, bringing an output value of CNY 2.444 million based on an average price of CNY 12/kg. The total output value was CNY 3.09 million.

(3) Cost input.

The total cost input for production, including labor, ducklings, feed, seeds, facilities, equipment, etc., was CNY 825,000.

(4) Economic benefits.

After subtracting the cost input from the planting and breeding output value of the base, the total income was CNY 2.265 million, and the income per hectare was CNY 61716, which was 4.6 times that of conventional planting.

III. Integrated Rice-*Trionyx* Farming Mode

[Case] Junshun Family Farm in Guozhuang Town (*Trionyx sinensis* and rice as the primary products)

In 2019, 1.67 hm^2 of the integrated farming area in the paddy fields of this family farm adopted the integrated rice-*Trionyx* farming mode.

(1) Farmland works.

Circular ditches were excavated on four sides. The ditch had an average width of 3.5 m with a water depth of 1.2 m, and the water surface accounts for 10% of the paddy field area. In the corner of the paddy field, there was a natural pond of about 200 m^2 as a temporary pond for breeding, with a depth of 2–3 m, and 15 m^2 of soil piled as a drying flat. Breeding density: 40 male *Trionyx sinensis* seedlings of 350 g size were bred per mu, and about 1,000 *Trionyx sinensis* were bred in total. Breeding management: Freshwater shrimp seedlings were placed in the circular ditches as natural bait, supplemented with chilled fish bait.

They were fed once every other day and about 30 kg each time. Harvest: The survival rate was more than 85%, and the average specification of each *Trionyx sinensis* at the end of 2020 was 1,500 g.

(2) Sales value.

In 2020, 450 commercial *Trionyx sinensis* weighing about 675 kg were sold, with a sales volume of CNY 135,000; the paddy field yielded rice 5,775 kg/hm^2, bringing an output value of CNY 106,087 based on an average price of CNY 11/kg. The total output value was CNY 241,087.

(3) Cost input.

The total cost input for production, including labor, *Trionyx sinensis* seedlings, feed, seeds, facilities, equipment, etc., was CNY 126,587.

(4) Economic benefits.

After subtracting the cost input from the planting and breeding output value of the base, the total income was CNY 114,500, and the income per hectare from paddy fields in the current season was CNY 68,563. In addition, in the paddy fields, there remained more than 400 *Trionyx sinesis* of the current season which were not ready for sale.

Section II Social and Economic Benefits

I. Social Benefits

First, the most important part of integrated farming in paddy fields is to make full use of land resources to achieve the purposes of "no conflict between grain and land" and "dual use of land and water", so as to maximize economic benefits and improve the economic income of farmers and fishermen. At the same time, integrated farming in paddy fields has realized the organic combination of planting and breeding, optimized the rural economic structure in the water network areas, promoted the full utilization of material and energy resources in paddy fields, broken the extensive production and operation mode of low-lying wetlands, and improved agricultural efficiency and farmers' income. From the perspective of the benefits of the integrated farming mode of paddy fields in recent years, the above farming modes of paddy fields have increased the net income by 3–5 times compared with the traditional rice-wheat planting mode, which shows great potential for promotion and application.

Second, no chemical pesticides and fertilizers are applied in the production process of integrated farming in paddy fields, which helps to improve the quality of agricultural products, ensure the safety of agricultural products, protect the environment and increase the benefits of rice planting. The agricultural products produced under the integrated farming mode of paddy fields will be safer and healthier. With the promotion of such modes, not only green and high-quality rice but also delicious poultry and aquatic products have been produced. Integrated farming in paddy fields has the advantages of low resource consumption, little environmental pollution, high product quality, and good economic benefits, and this meets the requirements of current social development for agricultural production. It has optimized the agricultural structure, improved the quality and safety of agricultural products, enhanced the market competitiveness of agricultural products, increased the market value of agricultural products, and realized the coordination and unity of resources and environment, quality, and benefits. Its characteristics of being green and environment-friendly have effectively driven the increase of economic benefits and to a certain extent, improved farmers' enthusiasm for farming.

Third, integrated farming in paddy fields helps to improve the comprehensive quality of grain farmers. Through the construction of the demonstration site of integrated farming in paddy fields, various technical training and on-site observation have been organized for farmers, which has improved farmers' theoretical knowledge level, availed them of an intuitive understanding of new things, and reserved a large number of reserve forces for the sustainable development of agriculture.

Fourthly, integrated farming in paddy fields has advanced agricultural industrialization. With the continuous promotion of green cultivation modes including integrated farming in paddy fields, the large-scale, standardized, and branded development of rice, aquatic products, and other industries has advanced. Based on the market demand, through information sharing, measures adapted to local conditions, the full play of local characteristics and advantages, the elimination of innumerable, disordered and confused brands, and focus on building a unified brand and improving the level, famous and popular brands come out; by further promotion of the centralized and contiguous farming of the same local mode and production of high-quality rice and aquaculture products with brand competitiveness, regional brand characteristics have been shaped. At the same time, the implementation of "dual use and dual harvest of the same paddy field" has greatly improved the output rate of cultivated land, promoted land transfer, derived a number of agriculture-related enterprises, family farms, and specialized agricultural cooperatives, and promoted the adjustment of agricultural industrial structure.

II. Ecological Benefits

First, it protects the ecological environment and water resources of farmland. For one thing, intercropping organisms in paddy fields can feed on weeds and insects, thereby lessening the damage of weeds and pests and reducing the use of pesticides; for another thing, the metabolites produced by intercropping organisms are absorbed by paddy field soil as high-quality organic fertilizer, which improves soil fertility and reduces the use of chemical fertilizers. The application amount of chemical fertilizers and pesticides can be reduced by more than 50% in fields that adopt integrated farming, which effectively reduces the adverse effects of chemical fertilizers and pesticides on the environment, protects the ecological environment of farmland, and at the same time realizes the recycling of water resources and ensures that water resources are free from pollution.

Second, it helps to improve the rural environment. In some areas of China, high input and high consumption during agricultural production have resulted in rural ecological degradation, environmental deterioration, and a poor rural living environment. By promoting the integrated farming mode of paddy fields, which requires scientific and rational allocation of various resources based on following natural laws, the goal is to maximize input-output benefits and optimize the ecological environment. This approach aims to restore the original appearance of the countryside characterized by "blue sky, clear water, and pure land".

Chapter VI Extension and Application of Technology

Section I Overview of Agricultural Science and Technology Extension

Agricultural extension is a social undertaking that occurs and develops along with agricultural production activities and is also in the service of agricultural production activities. The concept of "agricultural extension" had been used before the founding of the People's Republic of China. After the founding of the People's Republic of China, it was often referred to as "agricultural technology extension", and now it is gradually adopted as "agricultural science and technology extension".

I. Agricultural Science and Technology Extension & Agricultural Science and Technology Service

In a narrow sense, agricultural science and technology extension refers to the process by which agricultural scientific research institutes' achievements are promoted and popularized by extension personnel, enabling farmers to master these technologies and convert them into tangible productivity, thereby improving their agricultural income. The *Agricultural Technology Extension Law,* adopted by China on July 2, 1993, clarified that agricultural extension should cover the entire agricultural production process. It states that State-established agricultural extension institutions should popularize and demonstrate key technologies needed by the public with the support of agricultural technology experimental demonstration bases, providing corresponding non-profit agricultural technical services to farmers and agricultural production and operation organizations. In a broad sense, agricultural extension refers to services provided within rural society and for farmers, centered on farms or farmers, focusing on the promotion and communication of new technologies, new equipment, new concepts, and new information, and aiming at healthy and sustainable agricultural production. Agricultural extension activities, in a broad sense, are not only about disseminating new technologies but also involve organized and planned social activities, emphasizing the process of social education and publicity. In general, agricultural science and technology extension refers to a kind of social activity that disseminates new technologies, new concepts, and new information to farmers through demonstration, training, exchange, and other means of communication, and guides farmers to apply new technologies to production practice, so as to improve the quality, efficiency, and income of agriculture.

Agricultural science and technology service encompasses scientific and technological service activities, including consulting, transferring, and promoting scientific and technological achievements. These services are carried out by agricultural institutions under government administrative departments, such as scientific research institutes and professional associations, as well as by enterprises and individuals. The services provided include the introduction of new scientific and technological achievements, the application guidance of agricultural technological achievements, the establishment of scientific and technological consultation and guidance stations, the training of rural scientific and technological personnel, the training of farmers, and the continuous improvement of technical measures according to the needs of farmers.

Agricultural science and technology service includes two meanings: public welfare and marketability. The former requires that agricultural science and technology research and development (R&D) and extension personnel provide explanations and guidance to farmers in the theory of technical and scientific achievements and the practical operation of technology through various channels, while the latter requires agriculture-related enterprises, farmers' cooperatives and farmers to provide corresponding services according to the scientific and technological needs of agriculture created by production and rural social development. As an important component of the national public service, agricultural science and technology service is an activity in which agricultural enterprises actively engage, collaborating with agricultural research institutes and operating under the government leadership. It is a crucial part of agricultural productive services, focusing on helping farmers in conducting agricultural production activities through scientific and technological services and applications. This support aims to help farmers increase their income and achieve sustainable development of the rural economy.

II. Foreign Mode of Agricultural Science and Technology Extension and Service

The foreign mode of agricultural science and technology extension and service could be traced back to the Renaissance in Europe. During the 16th and 17th centuries, with the emergence of modern science, it was necessary to apply scientific achievements to meet the needs of human production and life, and thus early agricultural education began. Developed countries started the process of agricultural modernization early. They have formed their own distinctive science and technology extension service systems depending on the national conditions of the country, and their relevant research results are comprehensive, systematic, and in-depth. According to different extension subjects, foreign agricultural technology extension can be categorized into the following modes.

① Government-led mode: Agricultural scientific research and extension activities are under the organization and management of the government, and this mode is generally established by the agricultural technology extension station under the agricultural management department in conjunction with scientific research institutes and universities. It features vertical management and relies on the government's dominance in material, financial, and influential resources, as seen in Israel, South Korea, and Canada.

② University-led mode: The university-led extension mode, typically represented by the land-grant college mode in the United States, is also known as the "trinity" cooperative agricultural extension system of education, scientific research and promotion. The Government of the United States provides land for free, and the income from such land is used as university funds to maintain and fund the university's scientific research. Agricultural extension stations are set up in state universities to provide guidance on crop production to grassroots farmers in the fields. Each state has established agricultural experimental stations, and combined them with the land-grant colleges, forming a "trinity" cooperative agricultural extension system of education, scientific research, and promotion with universities as the core. This ensures the fairness and scientificity of agricultural extension while also facilitating a high adoption rate of new technological achievements, reflecting the principle of public welfare.

③ Division and collaboration mode: Represented by the Netherlands, this mode features a developed, efficient, and coordinated agricultural knowledge innovation and communication system, which is an important source of sustainable competitive advantage. This system focuses on establishing extensive and close links and interfaces with producers and operators, known as the Dutch "trinity" system for the coordinated operation of scientific research, promotion, and education. Various research institutions generate agricultural knowledge through research, and agricultural extension stations and industrial associations spread new agricultural knowledge and technologies to producers, and operators through education, demonstration, technical guidance, consultation, and other forms, and provide them with various agricultural productive services. In addition, various educational institutions improve the education level of family farmers and other agricultural practitioners through training and education, so as to promote the further communication of agricultural knowledge and technology.

④ Cooperative-led mode: Taking agricultural cooperatives as the carrier of the transformation and extension of scientific and technological achievements, it is the most important form of agricultural science and technology extension system. In European countries, Where small- and medium-sized farms are prevalent, countries like France rely on cooperatives for agricultural science and technology extension. Basic research projects are still government-led, while extension activities are conducted by agricultural cooperatives. In Japan, a combination of government and private sector efforts ensures timely collection of production issues and rapid conversion of research into practical applications, greatly improving the efficiency of agricultural science and technology extension.

III. Agricultural Science and Technology Extension and Service Mode in China

In the context of China's rapid advancement in agricultural modernization, the modes of agricultural technology extension and service have gradually become diversified.

1. Government-led mode

Taking the agricultural technology extension station as the main body, this mode belongs to the public welfare agricultural technology extension. The extension of new varieties and new technologies is mainly carried out through the combination of project campaigns, demonstrations, technology innovation and information consultation, product sales, after-sales service, etc. Due to the characteristics of linkage between the upper and lower levels, great motivation, wide popularization, fast promotion, and obvious effect, this mode has played a crucial role in China's agricultural technology innovation and achievement transformation. With the advent of the information age, in order to speed up the transformation of scientific and technological achievements, China has provided training for grassroots agricultural technicians, encouraged the use of the Internet and other modern information technology means, built agricultural technology promotion information platforms, and has shortened the intermediate links of agricultural technology extension, so that the efficiency of agricultural technology extension has been improved comprehensively. In addition, "Agricultural Technology 110", science and technology correspondents, and other government-led service modes are also adopted in some areas of China.

2. University-based mode

In recent years, China's colleges and universities of agriculture and forestry have accumulated rich experience in agricultural science and technology extension through exploration and practice, and formed a university-based agricultural extension mode. According to local characteristics and industrial needs, local governments cooperate with scientific research institutes, forming various extension modes such as "agricultural research institutes + local government departments", "agricultural research institutes + agricultural enterprises", and "agricultural research institutes + agricultural material production enterprises". Since the 1990s, Nanjing Agricultural University has carried out the "Science and Technology Caravan" (a touring show of agricultural technology and knowledge) and the "Double Hundred Project" (a project of prospering well off in one hundred villages through science and education by one hundred professors). By 2015, it had undertaken the agricultural extension service pilot work of the Ministry of Agriculture and the Ministry of Finance, and made innovations in the chain agricultural technology extension service mode of "scientific research and test base + regional demonstration base + grassroots extension service system + new agricultural business entity". The Northwest A&F University and the Baoji Municipal Government cooperated in 2000 to explore a "agricultural science and technology expert compound" mode that would meet the needs of local industrial development, so as to achieve "zero-distance" communication between experts, professors, and farmers. The mode is based on the local leading industry, with the local government providing support in the venue, capital, and policy, experts providing agricultural scientific and technological achievements, and agricultural producers participating in the extension and transformation of the achievements. In 2009, Professor Zhang Fusuo from China Agricultural University and his team established a "Science and Technology Backyard" in Baizhai Township, Quzhou County, Hebei Province, where they carried out scientific research and social services and achieved perfect results. Hebei Agricultural University has created a cross-departmental and cross-unit cooperation mechanism based on university-local government cooperation, university-enterprise cooperation, and base construction, with major science and technology projects as the carrier, and innovated the industry-university-research cooperation mode, which is known as the "Taihang Mountain Road" (representing the spirit of fearlessness in the face of hardships and difficulties) mode.

With the implementation of the rural revitalization strategy, agriculture-related colleges and universities have kept pace with the times, given full play to the role of "teaching" and "research", accurately interfaced with "agriculture, rural areas, and farmers", contributed to local development, and gradually played a professional advantage in the construction of university-local government and university-enterprise collaborative innovation platforms.

3. Academy of Agricultural Sciences-based mode

It is an agricultural extension mode led by agricultural science institutes of all levels, which has many advantages in science and technology, information, talents, practice, etc. The agricultural scientific research institutions under the Ministry of Agriculture and Rural Affairs and the provincial government are mainly responsible for basic research and applied basic research. The prefecture-level agricultural science institutes are mainly responsible for strengthening agricultural scientific and technological innovation, providing robust agricultural scientific and technological services, actively participating in agricultural science and technology extension, and aiming to promote the development of the agricultural economy in the local region. For example, the new agricultural technology promotion mode of Jiangsu Academy of Agricultural Sciences, with retired agricultural experts as the main body, makes full use of the advantages of retired expert groups with strong professional skills, abundant time and relatively low promotion cost, and is welcomed by the served area, winning the reputation of farmers and generating certain economic and social benefits. Relying on its own advantages in science and technology, the Guangdong Academy of Agricultural Sciences has built a "trinity" government-research-enterprise agricultural technology extension mode jointly with enterprises, so that the stock of grassroots agricultural technology promotion can be activated by the increment of science and technology. Hubei Academy of Agricultural Sciences has explored the "technical general contracting service" mode oriented by industrial needs, and signs technical service contracts with enterprises and governments to carry out targeted scientific and technological innovation and technical services.

The advantages of extension mode led by scientific research institutions such as universities and academies of agricultural sciences include high flexibility and timeliness, strong locality and applicability of technology research and development, and various forms of cooperation, while the limitations include farmers' lack of awareness of production behavior reform, dependence on continuous funding, and differences between the main assessment mechanism and the extension service orientation.

4. Enterprise-led mode

Through the integrated utilization of capital, technology, talents, and other factors of production, agriculture-related enterprises drive farmers to carry out specialized, standardized, and large-scale production, thus forming technology extension and operation modes such as "enterprise + base + farmer" and "enterprise + scientific and technological unit + farmer". This enterprise-led mode can increase both farmers' income and enterprises' profits. Enterprises play an important role in the development of agricultural production, technological innovation, as well as technological application and extension for its capability in flexible reflection of market information, full and reasonable mobilization of advantageous resources of all parties, and guiding enterprises' production and operation. However, this mode may also be restricted by factors such as weak scientific research ability, agricultural extension costs being greatly affected by the scale and condition of enterprise operation, and shortage of agricultural extension personnel within enterprises.

The enterprise-led mode has the advantages of creating an interest linkage mechanism with farmers and is characterized by its flexible and efficient. However, its limitations include a lack of user trust, barriers to technological innovation, and insufficient policy support for enterprises.

5. Cooperative-led mode

The cooperative-led mode is led by cooperative farmers and guided by the government. Professional big farmers take the lead, and smallholder farmers join, which is the way how it works. The information obtained comes from information sharing within the cooperative, and cooperative members are highly motivated to participate in the production. The cooperative-led agricultural technology extension mode fundamentally stimulates farmers' production enthusiasm, increases production flexibility and enhances production diversity, which contributes to the production and income increase of farmers and plays a greater role in promoting the further development of current agriculture. The cooperative-led mode of agricultural science and technology extension, which is characterized by strong inclusiveness, wide platform scope, and fast information dissemination and diffusion, can provide various technical services for farmers in agricultural production and promote the spread of new varieties.

The cooperative-led mode, as the main mode in the spontaneous modes of farmers, has an advantage in strong demonstration effect, while its limitations lie in limited coverage and insufficient technical support.

Section II Agricultural Extension and Operation Services

I. Overview of Agricultural Extension and Operation Services

1. Guiding ideology of agricultural extension and operation services

"Service as the purpose, technology as the core, operation for development" is the guiding ideology of agricultural extension and operation services. Agricultural extension and operation services aim to solve various practical problems in farmers' production and life by providing comprehensive, full-process service from agricultural extension institutions, so as to ensure the normal operation of agricultural production, realize the optimal combination of production factors, and obtain the best benefits.

Through operation services, the strength and vitality of the promotion agencies will be enhanced, the working and living treatment of the promotion personnel will be improved, the promotion teams will be stabilized and developed, and the development of agricultural extension will be promoted.

2. Basic principles of agricultural extension and operation services

(1) Adhere to the principle of combining "software" and "hardware".

(2) Adhere to the principle of profitability.

(3) Adhere to the principle of free trade.

(4) Adhere to the principle of conforming to regional industrial development.

(5) Adhere to the principle of observing disciplines and laws.

3. Business scope of agricultural extension and operation services

The business scope of agricultural extension and operation services includes providing information and material services before production, providing technical services during production, and providing storage, transportation, processing, and sales services after production.

4. Types of extension and operation services

(1) Operation services of economic entity of scientific research.

In the era of planned economy, agricultural research and extension were distinct entities. With the establishment of market economy, many business entities integrating scientific research, production, and operation services have emerged in recent years such as Denghai Seed Industry Group, Jinhai Seed Industry in Shandong Province, and Shandong Agricultural Fertilizer Technology Co., Ltd. These economic entities process technical achievements, operational capabilities, and the functions of self-accumulation and self-development, representing a future development trend.

(2) Production-supply-marketing integrated operation entity services.

Traditional services of agricultural extension institutions focus on providing technical services during production, while the production-supply-marketing integrated entity integrates production, supply, and distribution, which can not only supply the main production materials, but also provide technical services, which is conducive to agricultural production and self-accumulation of companies. For example, in terms of seed companies and feed companies at all levels, although they do not have technical achievements with independent intellectual property, they can purchase technical achievements through technology transfer and establish (or transfer in) their own production bases, focusing on production and operation.

(3) Technology-integrated operation service.

Entities of this nature lack their own achievements or production bases. It is mainly based on various distribution services carried out in combination with its own work of free services, from which profits can be made to subsidize the insufficient extension funds. For example, the crop protection station, soil and fertilizer station, forestry station, aquatic product station, crop hospital, animal hospital, aquatic production supplies service department, etc. under the Agricultural and Rural Bureau of the county.

(4) Group's contracted operation services.

Such services are mainly contracted services carried out by surplus personnel of agricultural technology extension institutions under the principle of voluntariness, such as co-production contracting, yield contracting, operation contracting, etc. Such operations are more common in the early stage of reform. At present, due to remuneration, not easy to define the effect of technology, and other reasons, such operations are rare at present.

(5) Intermediary technical services.

This kind of operation is the technology market, which belongs to the higher level technology operation. It mainly utilizes information advantages to provide intermediary services for holders of agricultural scientific and technological achievements, as well as agricultural technology demanders and users. This type of operation is most needed at present, but it is not perfect.

In addition to the first and fifth categories that mentioned above, the other three categories are all subordinate enterprises of agricultural extension agencies. For example, although the seed companies in each county have the status of legal persons, conduct independent accounting, and operate independently, they are under the leadership of the Agriculture and Rural Bureau, and most of the incomes are handed over. At present, they have not been completely decoupled from the parent company.

5. Marketing concept of agricultural extension and operation services

User concept, quality concept, service concept, value concept, benefit concept, competition concept, innovation concept, information concept, timeliness concept, and strategic concept.

II. Procedures for Agricultural Extension and Operation Services

1. Keep abreast of laws, regulations, and policies on agriculture

(1) Laws, regulations, and policies on agriculture and rural areas.

(2) Policies on agricultural material supply and services.

(3) Policies on credit and taxation of rural areas.

2. Carefully analyze the market environment

When analyzing the market environment, it is necessary to fully consider population, economic factors, competitive factors, scientific and technological factors, political factors, cultural factors, etc.

3. Carefully analyze the types and characteristics of farmers' purchase motivations

Seek innovation, novelty, beauty, brand, common ground, truth, integrity, and practice.

4. Identify target markets

To operate a service, it is necessary to consider who the buyer is; that is, take the buyer as a market. Different buyers have different personalities, hobbies, purchasing power, and purchasing purposes. There is a certain difference in demand, which is manifested as diversification of demand. Market segmentation is to classify buyers with similar needs into one category.

Operation service providers organize specific promotion projects and supporting measures to occupy a large market share and achieve the best operation effect according to the needs of the market segment.

There are three steps to determining the target market: predicting the demand of the target market, analyzing one's own competitive advantages, and selecting the strategic positioning of the market, i.e. precise prediction, fill-the-gap, and being innovative.

5. Determine the overall marketing mix strategy

The marketing mix of agricultural extension, namely the organic combination of strategy and tactics of agricultural extension marketing, is an important concept in the theoretical system of marketing.

The competitive position and characteristics of operators in the target market are fully reflected through the characteristics of the marketing mix.

4P strategy of marketing: product, place, promotion, and price.

6. Operation decision-making procedures and implementation plans

① Find problems and determine decision-making objectives; ② Formulate various alternatives; ③ Evaluate and select schemes; ④ Implement decision-making schemes.

III. Agricultural Extension and Marketing Skills

1. Occupy the market with high-quality products

The American scholars put forward the three-level theory of products; that is, any product can be theoretically divided into three levels: core product, actual product, and augmented product. The core product is a problem-solving service that explains the actual reasons why consumers buy or use the product. The core level of the product should be able to help users solve the most basic problems. For example, the core of food is to satisfy the needs of hunger and nutrition. The actual layer of a product is a visible and tangible level of product that transforms a product into a tangible entity or service. This level has five characteristics: quality level, features, brand name, styling, and packaging. This is the most intuitive and attractive layer for users. If actual product is a service, it should have similar characteristics. The augmented layer of the product means that the manufacturer can provide consumers with more services and benefits than physical goods. Examples include free installation, maintenance service, etc. This concept requires marketers to think more about providing services that exceed consumers' expectations and giving them a sense of complete satisfaction, so as to improve consumers' satisfaction and repurchase rate.

In the promotion of integrated farming products, we should provide a marketable product portfolio according to market demand, win by the quality and win consumer trust, carefully design product packaging, and establish brand image.

2. Flexible use of price competition to improve operating efficiency

The main strategies of price competition include discount pricing, regional pricing, and differential pricing.

3. Increase advertising and actively explore the market

The basic steps of advertising include formulating advertising plans, selecting appropriate advertising strategies, designing advertising, and specifying product details, as well as choosing the right advertising media. Commonly used advertising media include news advertisements, outdoor advertisements, store advertisements, traffic advertisements, entertainment advertisements, mail advertisements, gift advertisements, exhibition advertisements, sample description advertisements, and decoration advertisements.

4. Utilize modern promotion skills to promote products

(1) The elements of promotion include means, principles, modes, and methods.

① Means of promotion: evidence, reason, and reputation.

② Principles of promotion: user-centric, reality-based, and effect.

③ Modes of promotion: push and reverse attraction.

④ Methods of promotion: personal selling, advertising, business promotion, public relations, and special promotion.

(2) Promotion skills.

① Interview skills.

② Skills for handling objections.

③ Skills of persuasion and advice.

④ Repeatability and scheduling skills.

⑤ Negotiation skills.

(3) Cultivate skilled salespersons.

Salespersons are trained in professional skills and public relations skills.

Section III Promotion Cases of Integrated Rice-Fishery Farming

I. Integrated Rice-Fishery Farming in China

According to the *Development Report of Integrated Rice-Fishery Farming Industry in China* (2018), there were 27 provinces with integrated rice-fishery farming in China in 2017, but no statistics for Beijing, Hainan, Tibet, and Qinghai were found, with a total area of 28 million mu, including 5.02 million mu in Hubei Province, 4.64 million mu in Sichuan Province, and 3.32 million mu in Hunan Province. The total area of the above three provinces accounts for 46.4% of the total area of integrated rice-fishery farming in China. In addition, the total area of integrated rice-fishery farming in Jiangsu, Guizhou, Yunnan, Anhui, and Zhejiang exceeds more than 1 million mu.

In Yunnan, Jiangxi, Hunan, Hubei, and Chongqing, traditional manual-transplanting techniques are mainly used in paddy fields, including fish farming, snail farming, and crayfish farming. Jiangsu Province mainly uses advanced machine-transplanting techniques in paddy fields, including duck farming, fish farming, *Trionyx* farming, and crayfish farming. Heilongjiang and other northeastern provinces also use machine-transplanting techniques in paddy fields, including duck farming, fish farming, *Trionyx* farming, and crayfish farming. Loach farming, crab farming, and leech farming are also used in paddy fields, while in small areas, base farming continues to be the dominant practice.

II. Development in Jiangsu Province

With the support of relevant policies and projects, the integrated farming area of paddy fields in Jiangsu Province has developed rapidly, mainly concentrated in northern Jiangsu, of which Jiangyan Base in Taizhou is its typical representative. In recent years, all parts of the province have continuously innovated technology, optimized varieties, and actively promoted integrated farming modes, such as rice-crayfish, rice-crab, and rice-duck. The integrated farming area of paddy fields in Jiangsu increased from 100,000 mu in 2016 to 1.5 million mu in 2019 and shown an upward trend. After preliminary understanding, more than 30 counties (cities and districts) in the province have issued incentive policies and have done a lot of work in integrated farming innovation and technology demonstration. Xuyi County, Pei County, Sihong County, etc. were awarded the title of "National Integrated Rice-Fishery Farming Demonstration Area" by the Ministry of Agriculture and Rural Affairs of the People's Republic of China.

III. Application Cases in Jurong in Southern Jiangsu

Located in the south of Jiangsu Province, Jurong is one of the top 100 counties in China and also one of the top 100 counties and cities in China for green development. It borders the Yangtze River in the North and is the birthplace of the Qinhuai River. It has rich forest resources, with Baohua Mountain in the North and Maoshan Mountain in the South. It is known as "five mountains, one river, and four fields", which is a typical hilly area. The cultivated area is about 53,300 hm^2, and the perennial sown area of crops is about 76,700 hm^2. It mainly grows grain and oil crops such as rice, wheat, and rape, as well as fresh fruits such as grape, strawberry, pear, and peach, with a total agricultural output value of more than CNY 4 billion.

The perennial planting area of rice is about 25,300 hm^2. In recent years, with the promotion of palatable rice, the planting scale of palatable rice varieties in the jurisdiction has gradually expanded, and the planted Nanjing series rice has become more and more popular in the market of southern Jiangsu with its excellent flavor. At the same time, the planting scale of organic green rice in the jurisdiction continues to increase, and the scale of agricultural enterprises engaged in rice production, processing, and sales is gradually expanding.

In the current environment, characterized by the low output value of grain crops, higher costs, and the shrinking benefits of agricultural production, the previous model of relying on large-scale planting to increase income through high yields has significant shortcomings. The development of green, circular, high-quality, and efficient characteristic agriculture is conducive to reducing production costs, achieving the goals of good quality, safety, and security of agricultural products, high utilization of resources, and high production benefits through the application of safe and efficient agricultural new technologies and modes, and truly realizing the transformation of agricultural production orientation from growth to quality improvement.

1. At the beginning

At the end of the 1990s, integrated farming of paddy fields began to develop sporadically in Jurong. At first, the Houbai Good Seeds Farm implemented integrated rice-duck farming with the support and help of the Zhenjiang Science and Technology Commission (now Zhenjiang Science and Technology Bureau). In the later stage, the farming area reached 3.33 hm^2. The duck was the working duck, 15–20 ducks were stocked every mu, and the mature duck was about 1.2–15 kg. Most of the ducks were sold in vacuum-packaged sauce ducks. However, due to the lack of full recognition of high-quality agricultural products in the market at that time, most of the rice and ducks produced were sold at the prices of ordinary agricultural products after entering the market, without achieving high quality and good price, and the production benefits were not high. After several years of development, the area of this kind of farming mode continued to shrink and then disappeared.

2. Development status

In recent years, high-quality agricultural products such as pollution-free and organic green products have become more and more popular in the market. The area of integrated farming of paddy fields in Jurong has been continuously expanded, showing new characteristics. ① Diversified modes: The mode develops from the original integrated rice-duck farming to rice-crayfish, rice-*Trionyx*, rice-frog, and other integrated farming modes. ② Large-scale production: The implementation subject develops from small-scale and scattered ordinary farmers to new business entities with large implementation scales. ③ Advancement of technology: Rice varieties have changed from varieties with high-production to palatable varieties, relevant input products have also changed from chemical fertilizers and chemical pesticides to green and environmentally friendly products, and field management and feeding technology are also more scientific.

3. Achievements

There are many organic and green-certified rice brands in Jurong, such as Yueguang Rice, Du's Tianxiang, and Lvyuan Rice, which are widely praised for safety, high quality, and palatable taste. At present, the area of integrated rice-duck farming is the largest in Jurong, among which the technology of rice-duck and green manure is the most prominent. The specific operation process is as follows: In mid- and late October, namely about one week before rice harvest, green manure is interplanted in the paddy field, and the green manure varieties are mainly *Astragalus sinicus* L. The rice is harvested in due course. Next year, in mid-of April, green manure is plowed; rice seedlings are raised in early and mid-May; rice is planted in the first half of June; ducks are released in the middle and lower half June; ducks are withdrawn from paddy field before rice heading. The whole process is produced in a green and organic cutting method. Benefiting from the improvement of rice quality through agricultural techniques such as integrated farming of paddy fields, rice produced in Jurong is of good quality and has won many provincial and municipal competitions of rice. In the first Zhenjiang Good Rice Selection, it won 2 gold awards: Du's Tianxiang of Du's Family Farm in Houbai ranking first and Lvyuan Rice of Houbai Good Seeds Farm ranking fourth (there are a total of five gold awards in this selection).

4. Deficiencies

Currently, there is significant variation in the technical proficiency and management practices among different integrated farming bases in Jurong' s paddy fields. Consequently, ducks, fish, and crayfish have experienced varying degrees of mortality. During the process of integrating duck rearing with paddy cultivation, a lack of understanding of the technology has led some farmers to keep ducklings in pens for extended periods to minimize losses, rather than introducing them to the paddy fields once the rice seedlings are established. As a result, weeds in paddy fields were not cleaned by ducks in time, and there were many weeds in some fields. The real purpose of integrated farming of paddy fields is not achieved. At the same time, the quantity and quality of rice, ducks, crayfish, and other products produced vary greatly, and the sales path and final benefits of each base are also quite different. According to the survey, the net benefit per mu of integrated rice-duck farming in Jurong in 2017 was CNY 1,370–6,800, mainly due to the large difference in the selling price of rice produced by each base. Considering the large-scale production and market acceptance (the price of rice cultivated by integrated rice-duck farming is 10%–30% higher than the purchase price of common rice). The theoretical benefit of raising ducks per mu is about CNY 3,400 (including CNY 2,400 for rice and CNY 1,000 for ducks); the cost per mu is about CNY 1,300 (including CNY 550 for rice and CNY 750 for ducks); the theoretical net benefit per mu is about CNY 2,100. Integrated rice-fishery farming mode is still in its infancy in Jurong, the implementation area is also small, and the related technology needs to be improved.